Advance Praise for
The Primate Myth

"Illuminating and paradigm changing! Jonathan Leaf overturns decades of assumptions about our human nature and its origin. Realizing that perhaps we have become too complacent in our human story, he poses the empirically driven question, what really has been our path of evolution? Leaf challenges and sheds light on our social, political, and truly life-or-death human decision-making. In a culture of individualism, do we realize our capacity for original thought and courage? Leaf is a profound, rare generational thinker. You will rethink our times with a renewed curiosity for human nature and our evolutionary cousins."

—PROFESSOR LISA MILLER, Professor of Psychology, Columbia University
and *New York Times'* National Bestselling Author of *The Awakened Brain*

"It's easy to be cynical about human beings. This book is an antidote. Leaf thinks scientists have made too much of our evolutionary continuity with chimpanzees, which are horrifically violent. In fact, we act more like wolves and dolphins—more altruistic, more prosocial, more creative. Leaf rethinks the famous Milgram electric-shock experiments, the factors leading to suicide, how and why we seek status, and the origins of sexual jealousy. It's a provocative, wide-ranging book—a must-read for anyone trying to understand human nature."

—GEORGE MUSSER, Contributing Editor, Scientific American and author
of *Putting Ourselves Back in the Equation: Why Physicists Are Studying
Human Consciousness and AI to Unravel the Mysteries of the Universe*

"Jonathan Leaf is a brilliant amateur who can illuminate almost anything and who sometimes sees clearly what the experts miss. The experts who have told us that humans are essentially upgraded chimpanzees, he says, have missed a wealth of details that make us more similar to dolphins and whales. We share a remote ancestor with chimps, but once our forebears left the canopy of the tropical forest, we evolved into a large-brained, pair-bonding, cooperative, organizing, and language speaking species. The key is that we became a special kind of herd animal. Leaf fearlessly follows this thread through unexpected territory including why we fight wars; why men are more likely than women to be homosexual; and why humans (but not chimps) commit suicide. The book reflects his deep reading in a dozen disciplines, but Leaf's career as a successful dramatist serves him well. Every page is alive with anecdotes, recollections, and stories of the people who enthroned chimpanzees as our close cousins and thus diverted attention from who we really are."

—PETER W. WOOD, President, National Association of Scholars and Former Anthropology Professor at Haverford and University of Rochester

"This book shreds conventional wisdom about the origins of human nature. Leaf is a compelling writer and, whichever side of the argument you come down on, you will emerge with a clearer understanding of the challenges facing humanity in an age of ubiquitous information and global aggression."

—PHILIP HOWARD, Author of *The Death of Common Sense*

"Jonathan Leaf delves into our brains and our prehistory to discover who we are: animals that hunt, speak, and follow the leader. Moving at ease through a mass of data, *The Primate Myth* is fascinating, occasionally contentious, and sometimes frightening."

—RICHARD BROOKHISER, Contributing Editor, *National Review*

"No one can come away from Jonathan Leaf's extraordinary survey of the twists and turns of what we know about human nature without recognizing how our understanding of human nature has been abused and distorted for agenda-driven causes. *The Primate Myth* should be required reading for generalists and specialists alike."

—STEVEN HAYWARD, Senior Resident Scholar, Institute of Governmental Studies, University of California, Berkeley

Also by Jonathan Leaf

City of Angles

THE PRIMATE MYTH

Why the Latest Science Leads Us to a New Theory of Human Nature

JONATHAN LEAF

BOMBARDIER
BOOKS

Published by Bombardier Books
An Imprint of Post Hill Press
ISBN: 979-8-89565-077-6
ISBN (eBook): 979-8-89565-078-3

The Primate Myth:
Why the Latest Science Leads Us to a New Theory of Human Nature
© 2025 by Jonathan Leaf
All Rights Reserved

Cover Design by Christina Freyss

BOMBARDIER
BOOKS

Post Hill
PRESS

Post Hill Press
New York • Nashville
posthillpress.com

Published in the United States of America
1 2 3 4 5 6 7 8 9 10

To my beloved wife.

~

"All truth passes through three stages: First, it is ridiculed; Second, it is violently opposed; and Third, it is accepted as self-evident."

—ARTHUR SCHOPENHAUER

~

THE PRIMATE MYTH

Primates are mammals with prehensile feet. These are used for climbing trees. Although we do not have them, we are classified as primates. Is this leading us to wrong ideas about basic human activities like sex, sports, fashion, suicide, and war? This chart shows how alike our nature is to one herd and one pack animal – and chimps.

SPECIES	HUMAN	BOTTLENOSE DOLPHIN	WOLF	CHIMPANZEE
Encephalization Quotient (Brain to Body Size Ratio)	High	High	Moderate	40% as big as dolphin brain and 30% of human
Carnivore	Yes – over half of pre-historic diet	Yes	Yes	Meat is just 1-3% of diet
Eats Carrion/ Scavenger	Yes	Yes	Yes	Never
Altruistic	Yes	Yes	Yes	Rarely, if ever
Tame	Yes	Yes	Partially	Extremely dangerous
Names Offspring/Uses Spoken Language	Yes	Yes	No	No, and is far inferior to dogs at learning words
Highly Cooperative	Yes	Yes	Yes	No
Arboreal/Tree Dweller	No	No	No	Yes
Endurance	Exceptional, lacks explosive strength	Exceptional	Exceptional	Poor, but has explosive strength
Capable of Co-Feeding	Displays instinct from earliest ages	Displays instinct from earliest ages	Displays instinct from earliest ages	Never shares food with unfamiliar chimps
Sings	Yes – often part of mating ritual	Yes – often part of mating ritual	Yes – often part of mating ritual	No
Eats Own Young/ Filial Cannibalism	Never	Never	Never	Frequently
Commits Suicide	Yes, individually and collectively	Yes, individually and collectively	No	Never
Characteristic Mating Pattern	Monogamous, lowest false paternity rate of any mammal studied	Polygamous	Monogamous, mating for life is common	Polygamous with no pair-bonding or awareness of paternity
Average Daily Sleep Hours/Sleep Patterns	7 hours – mates sleep together	8 hours (4 for each hemisphere of brain)	7 hours – mates sleep together	10.5 hours, males never allowed within tree nests
Parenting Method	Co-parenting by mates and other relatives	Shared parenting by females of herd	Co-parenting by mates and other relatives	No co-parenting, very rare cases of shared parenting

The Primate Myth presents revelatory new research in neuroscience and biochemistry. In doing so, it presents a challenge. Will you accept that *homo sapiens* is a herd animal? If you can read this book and still believe you're a primate, then you will have proved one thing: you, especially, are not.

TABLE OF CONTENTS

CHAPTER 1

THE PRIMATE MYTH

THEOPHILUS PAINTER WAS A FUDDY-DUDDY. You can see that in a youthful snapshot, a picture taken when he was barely out of his teens. His hair is cut two-thirds of an inch above his ears, and he is wearing a tight, heavily starched three-piece suit. His left hand is firmly planted in a front trouser pocket, and his look is stern. His thin lips are sutured together, and his cobalt blue eyes project a humorless conviction. The face displayed in the sepia-toned print is that of someone who believes that he has a special destiny. Persistently ill as a child, he was homeschooled, then sent at age fifteen to a local college. There Painter was promptly recognized as the most gifted student in his class. Four years later, in 1908, he entered Yale as a graduate student. In the photograph, his expression declares his conviction that he will be at the center of great events.

He was right about this. Wherever he was, history was being made.

He had arrived at Yale with plans to study chemistry, but he soon found himself pulled in another direction. One of his professors had made the discovery that living tissue could be grown in cell cultures, and, intent on making a name for himself, he decided to make use of the technique. Switching over to the university's doctoral program in biology, he turned his attention to the fertilization of egg cells. Initially, he occupied himself with spiders. Next, he examined sea urchins and lizards. Their embryonic cells are minuscule, and the work was challenging. Yet through his invention of a special knife with multiple blades, he was able to prepare slides with clearer images of the smallest parts. The process allowed him to see what had been invisible to that point: the male sex chromosomes of a mammal. The first to reveal itself under magnification was that of an opossum.

But Painter was intent on learning about humans. Obtaining sperm cells in the form he desired was not simple, and the method he decided upon would be impermissible today. An acquaintance at the Texas State Lunatic Asylum sent him the testicles of three patients who had been placed under anesthesia and castrated for "excessive" self-abuse. Painter took these and employed an elaborate process by which he could section them and arrange tissue-thin slices for microscopic inspection. Then he photographed the precise drawings he had made of what he had seen.

What emerged from the darkroom was a series of printed sheets. On these were squiggly black lines on a blank white field. His attention gravitated toward a single pair in which one of the two lines was noticeably shorter than the other. Painter proceeded to make the bold suggestion that the abbreviated strip represented the chromosome coding for maleness: the Y chromosome. It was an extraordinary intuitive leap. Later scholars concluded that Painter might not have identified the right coupling. Yet his deduction that the Y chromosome was a dwarf version of the X chromosome to which it was attached was correct, and it brought him instant fame.

He had made a significant error though. While the resolution of his pictures was improved by his self-created knife, his photographs were still blurry, and he couldn't easily count up the number of chromosomes. Were there twenty-three pairs or twenty-four? Painter wasn't certain. He had also studied the sperm cells of rhesus monkeys, and he had shown that they had twenty-four pairs. Other scientists were reporting that chimps and gorillas had twenty-four pairs. Since these species are related to us it stood to reason that humans should have twenty-four too. Perhaps with this in mind, Painter finally settled on an answer, announcing to the world that humans had twenty-four pairs of chromosomes. Almost immediately this was accepted as a fact.

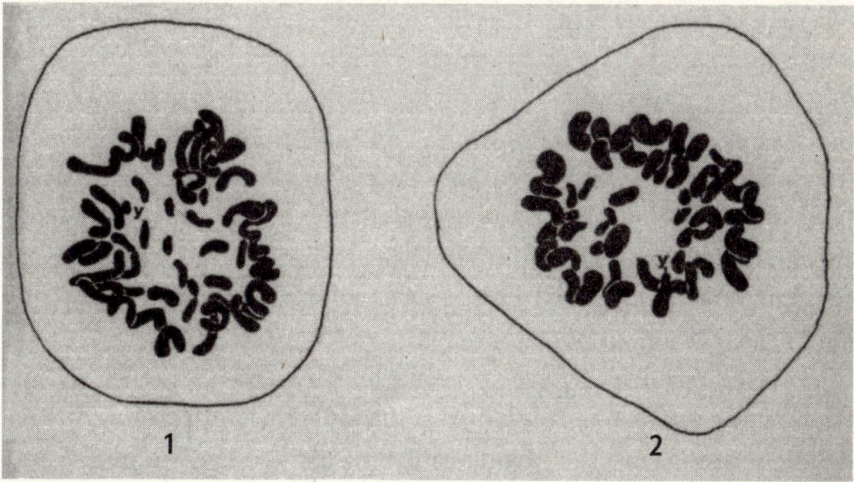

Theophilus Painter's slides of human chromosomes.

Over the following decades, thousands of scientists examined reproductions of Painter's most famous slides. Because the pictures were hazy, no one was able to say with any assurance if Painter's tally was accurate. Few bothered to do their own reckoning of what was shown in the images, and the technology for taking photographs enlarged from microscopic sections remained primitive. Thus, it was not until 1955—three decades after Painter had first spotted them— that Joe Hin Tjio, an Indonesian-born scientist living in Sweden, was able to do a proper recount. Tjio made use of improved staining agents. The new method provided sharper images, and, perusing his slides, Tjio realized to his astonishment that there were only twenty-three pairs—forty-six chromosomes.

In the weeks and months that followed, other scientists went back, eyeing the best-known image developed by Painter. Although it had been reproduced in millions of college and high school textbooks, to their amazement they realized that it could just as easily—perhaps even more easily—be interpreted to show twenty-three pairs with forty-six chromosomes. No one had caught the mistake.

In the meantime, Painter moved on to investigating the chromosomal structure of fruit flies. It was an ideal subject. As they have an unusually simple genetic structure and a life span of just a few weeks, they have been *the* means by which scientists have learned how genes function, and once again Painter was blessed not only with perfect timing but a useful and thoroughly novel technique for his research. It was based on the realization that radioactivity causes genetic mutations. Inducing these, he and a partner swiftly saw how genetic changes appeared and were passed on. That led to a series of scientific papers, written between 1929 and 1939, that revolutionized the understanding of the movement of genes between chromosome pairs.

Painter's academic celebrity brought a remarkable rise in his status. Starting out as a mere lecturer at the University of Texas at Austin, he was promoted all the way up to its president. That compelled him to confront new issues. For, throughout this period, the school was segregated. One of the many people who objected to this was a young African American named Heman Marion Sweatt.

Sweatt had an amount in common with Painter. Like him, he had been something of an intellectual prodigy, and he had acquired a broad knowledge of biology. Trained at the University of Michigan, where he was a medical student, Sweatt had particularly excelled in bacteriology and immunology. But, after dropping out, Sweatt had returned to Texas and gone to work in the same field that his father had. He was supporting himself as a mailman. Finding that even opportunities for promotion in the post office were denied to him, Sweatt became a civil rights activist and a columnist for a Black newspaper. Over time, he became increasingly aware that defeating Jim Crow required legal skills. So in 1946 he met with Painter to make a request: The University of Texas's law school should end its discriminatory admission practices, and he should become its first African American student. When Painter told him that this was impossible, Sweatt filed a class-action lawsuit naming Painter as the defendant. Sweatt's able lawyer, Thurgood Marshall Jr., brought the case to the US Supreme Court. Known today as *Sweatt v. Painter*, the lawsuit led directly to another case in which Marshall was the lead attorney: the *Brown v. Board of Education* decision that ended legal segregation.

Like many people of his era, Painter thought that Blacks and Whites must be different in their fundamental biology. So when he reported his results on chromosomes, he provided separate data on the two groups. The implication seemed to be that it was noteworthy, even unexpected, that Blacks and Whites had the same number. Perhaps not surprisingly then, Painter did not fight against the belief that Blacks and Whites should be kept apart. Dutifully, he obeyed the instructions given to him by the school's Board of Regents. These were to keep Sweatt out, and month after month, over full four years, he refused Sweatt's admission. Painter began by offering Sweatt scholarship money to attend an out-of-state law school. Then he tried to arrange for Sweatt to go to a newly constructed Blacks-only law school in Houston. This was to be housed within three basement rooms that would include a small law library. This contrasted with the law school in Austin, which had a whole building, sixteen full-time faculty members and 65,000 volumes. Nonetheless, Painter insisted that the school in Houston would be just as good.

These two beliefs—Painter's erroneous assertions about the number of human chromosomes and the notion that schools can be "separate but equal"—have a certain kinship. Both are examples of a phenomenon that is more commonplace than we might care to admit. Dubious opinions are offered up, and they live on for decades or even centuries. Often these false claims are easy to disprove. All that is required is dispassionate study. There are countless illustrations of this, cases in which absurd notions were accepted for long periods during which overwhelming proof of their error was ignored or rejected. Sometimes those who are the most mistaken are research scientists, men and women trained in the objective analysis of evidence.

Why is it that people become so committed to false beliefs? That riddle brings us to a larger enigma that Painter was investigating as he was peering into his microscope. *What is a man?* Attached to that is a second puzzle: *How did we become human?*

This book is not an attack on the theory of evolution. It does, however, question the claim that humans are sufficiently alike to chimps and gorillas that we should be placed in the same family and order of animals. Primate anthropologists are convinced that they have a vast amount to tell us about human nature. They insist that their research on chimps and gorillas is profoundly revealing about our basic modes of thought and behavior. But what if this is not so? Might it be like the idea that we have the same forty-eight chromosomes that chimps and gorillas have? Or that segregated schools will be equal? Is it also possible that there is something innate in humans making us unwilling to oppose or even susceptible to belief in widely held but mistaken ideas?

To answer those questions we need to know what primate researchers believe. Here are their basic propositions:

> *All primates have broad, flat nails and prehensile feet.*
> *The expression prehensile feet means that a primate's toes are designed for climbing trees.* But, although humans do not have these, we are primates.
> *Primates are arboreal.* Only a few primate species do not spend the majority of their time on tree branches, and all primates, except humans, spend some of their time there. This forest cover is to be found in the tropics or, more rarely, the subtropics. Primate existence is threatened by deforestation, a common practice that we employ to expand our domain, though we are supposedly primates.
> *Primates eat little meat or none whatever.*
> *There are five suborders of primates.* Two are comprised of the "twisted nose" creatures, like lemurs and bush babies. The other three are tarsiers, monkeys and apes. Apes are Old World primates that are without a monkey's tail.
> *The suborder of apes should be further split up into "lesser apes," or gibbons, and "great apes."* The category of great apes includes orangutans, gorillas, chimps, and humans.

> ➤ *Humans are greatly akin to chimps.* Indeed, although humans lack prehensile feet, and we are hairless, speaking, domesticated, pair-bonding, highly cooperative, meso-carnivorous bipeds who prefer to live on open plains, often far from the tropics, we are more like chimps than chimps are like gorillas. This is said to be so even though chimps and gorillas are both untamed, equatorial, forest-dwelling, herbivorous, tree-climbing knuckle-walkers who lack skills of speech and cooperativeness.

That last notion has been embraced not only in academia but in popular culture. The idea spread following the publication of Jane Goodall's 1971 book, *In the Shadow of Man.* A student of the renowned paleontologist Louis Leakey, Goodall had traveled from England to Tanzania twelve years earlier. Near Lake Tanganyika, she set up a field station in a three-and-a-half-square-mile stretch of tropical forest, what would come to be called the Gombe Stream National Park. Goodall's research, which continues to this day, focused on observation of chimpanzees in their natural setting, and her reports convinced people throughout the world that chimps behave much as humans do.

The currency of this notion increased immensely during the next three decades. As the average person does not read case studies detailing chimp life, what really popularized these beliefs was a series of successful movies and television shows. Foremost among these were *Planet of the Apes* and *Gorillas in the Mist.* These motion pictures promoted Goodall's view that humans were not so different from the great apes, and, as this idea gained general acceptance, she and other caretakers and researchers on ape species became adored heroines of animal lovers. Concurrently, at primate research laboratories experimenters began trying to teach young chimps human language and vocabulary, and the public was informed that they were ready and capable students—not so different from human children. This was also when the international bestseller *The Third Chimpanzee* was published. The work not of a primate expert but an ornithologist, it suggested that humans were just another species of chimp and that chimps should be classified as *Homo*: people, like us. As this notion reached its apogee, some families adopted chimps and tried to raise them as they would human children. There was even a prime-time network TV show about a family that took in a chimp, treating him like one of their own offspring.

The belief that humans and chimps are not so different is a subject of particular interest to college professors teaching philosophy. The ethicist Peter Singer and the evolutionary biologist Richard Dawkins have suggested that chimps—and apes generally—should be thought of as "non-human persons" and accorded the same rights and treatment as people. In 2008 this led Spain to express its support for the idea. Two years later the European Union declared that its scientists could not conduct any further research on apes. The sole exceptions permitted are when they face extinction and when humans are confronted by the "outbreak of a life-threatening or debilitating disease." The idea seems to be that as chimps and gorillas are our brothers and sisters, and as they have never placed their signatures on any piece of paper accepting experimentation upon them, it ought never to occur.

Throughout this period the general view was that apes were less violent than humans. Indeed, that apes killed one another was taken as proof that although they were essentially genial and benign, they had some of our savagery within them, and their occasional nastiness was seen as evidence of an underlying commonality. Yet even as the notion that chimps are our spiritual brethren was taken up by the greater portion of the intelligentsia, lawmakers, and much of the public, the evidence that this was not so was becoming increasingly hard to ignore.

This started with the experience of those who brought chimps into their homes. They found that they were not like cats or dogs. Frequently these adoptions ended in tragic acts of mayhem. People had their noses ripped off, their eyes gouged out, and their arms torn out of their sockets. Researchers were forced to acknowledge the obvious: Chimps aren't tame. Once they reach maturity, they become exceptionally dangerous. They do not take to human ways of life, and they are quite different from people.

In the early 1970s, it was claimed that chimps could master sign language and make whole sentences out of it. Those claims have been debunked. For all the assertions of chimp brilliance, it was eventually realized that they are far inferior to dogs at grasping speech and language. Research shows border collies can learn up to 1,000 words. Most studies of chimps have suggested that they cannot learn more than 250. They also learn more slowly than dogs. A good rate for a chimp is mastery of 2 new symbols in six months. However, a skilled trainer may be able to teach a dog 12 words in a week, and these are spoken words, not the symbols on a keyboard ordinarily taught to chimps. More remarkably, pattern-recognition tests reveal that chimps are no better at identifying and remembering sequences of flashing lights than monkeys and a number of species of birds—including chickens. Rats consistently outperform them.

Most primatologists remain convinced that chimps and gorillas are very much like people. So they are understandably uncomfortable discussing those facts. Intent on proving our kinship with chimps they will instead cite the results of what's known as the cooperative pulling experiment. This is meant to serve as proof of our commonality. In this test, animals work together to pull on a strap in order to receive a reward. Chimps can pass this test. What primate researchers neglect to mention is that wolves, dogs, hyenas, elephants, otters, parrots, crows, and ravens can also pass it—although none of them has proper hands. Primate researchers are also apt to point out that chimps hunt together. This ignores the obvious fact that countless species do, including falcons, lions, wolves, hyenas, and even ants.

Nor is their hunting all that frequent. No more than 3 percent of a *Pan troglodyte* chimp's diet is composed of meat. Some estimates place the amount as low as 1 percent. What's more, bonobos eat even less meat. In fact, one study that looked at blood proteins associated with meat consumption concluded that a small species of African antelope takes in proportionately more. However, the handful of observations presented by Goodall of chimp hunting had convinced a great many journalists and academics that chimps are hunters in much the way that people have been.

Research showing that chimps eat little meat has been ignored. A number of the best-known primatologists bear a measure of blame for this as some of them have deliberately obscured this point. The late Dutch primatologist Frans de Waal, for example, told his many readers that chimps eat "substantial amounts of meat, which they hunt themselves," purposefully but falsely implying that it has a central element in their diet and that they are adept and regular in their acquisition of it. When de Waal was named head of the Emory National Primate Research Center, he used his position to pressure the scientists working for him from presenting conclusions that might suggest humans are unlike chimps. In effect, he tried to quash the findings coming out of his laboratories that undermined his own theories. Those discoveries were based on neuroscience imaging. But, even before that technology appeared, primatologists had come to realize that something was wrong with the idea that humans are all that much like chimps.

One important clue turned up in 1975. That was when Jane Goodall returned to Tanzania from a brief trip to England. By that time, she had attracted a gaggle of aides. They told her about two chimps described in her book that had gone on a cannibalistic killing spree. It was the first but not the last time they would observe chimps consuming the flesh of other chimps. This practice went unreported in the mass media, and it has continued to be neither known nor understood for its implications by the general public. What had sunk in instead were cheery accounts of how chimps go about their labors. Thus, Goodall had remarked in wonderment and with considerable pride that chimps trying to capture colobus monkeys could act like people, engaging in "quite remarkable cooperation—as when different chimpanzees station themselves at the bases of trees offering escape routes to their cornered victims." A critical reading of this, of course, is that Goodall was telling her readers not that chimps are much like us but rather that they are capable on occasion of doing what wolves and hyenas do routinely. Only the attentive reader may have noticed what she also reported: Most often "the capture of a prey is accidental; the chimpanzee, as he wanders along, stumbles across a baby bushpig, grabs it, and the kill is made."

That view is ratified by a recent paper in *Nature* on chimp behavior which noted that "evidence of [chimp] communication in the context of coordination problems is scarce." In fact, more than a few field researchers have suggested that what Goodall was describing was uncharacteristic of chimps. They found that when chimps went out in larger groups, they hunted *less* effectively. The more chimps there were, the less meat each one got.

In general, chimps have little interest in working together, and they rarely display altruism. Indeed, when experimenters set up a contraption by which chimps had a choice of two ropes to pull and they could see that selecting one provided a meal to them and the other offered a meal to themselves and another chimp, it motivated them not at all. They "chose randomly." Even when helping another chimp out cost them nothing, they couldn't be bothered to do it. By contrast, human toddlers will regularly share food with friends and non-friends alike by age three. In fact, some studies show that nineteen-month-old babies like sharing food and they will give it to others when they are hungry themselves.

While Goodall was working in Tanzania, Frans de Waal was employed at a zoo in the Netherlands. His observations led him to the publication of his 1982 book, *Chimpanzee Politics: Power and Sex among Apes*. It told a different story. In de Waal's account, chimps were in constant competition and exceedingly violent, and this explained human brutality and aggression. Eventually de Waal went beyond this to suggest that chimp troops regularly undertook genocidal wars against each other. De Waal's claims were complicated by the fact that as a zookeeper he was studying captive chimps, not those in the wild. Even so, he had an enormous influence both within the field of anthropology and outside it.

In the meantime, Goodall and many others continued observing chimps in their native habitats. As Goodall loves the chimps she studies, in very human fashion she has devoted herself while aiming to see them in the best light. Yet what she and her assistants have witnessed does not jibe with the notion that they are much like us. Limited and circumscribed as their encounters have been, their accounts render us an astonishing record of atrocities. The more gruesome stories cannot be quickly forgotten. Soon after Goodall's book came out, her researchers observed two males that attacked females from outside their troop, stole their babies, tore the infants to shreds and digested them. More remarkable still was that cannibalistic mother-daughter team, known as Passion and Pom. Working side by side, they captured, killed, and consumed something in the range of ten chimp babies. Three seem to have been from their own troop. Overall, a study of the work of chimp researchers in Africa found that through 426 years of observations they had uncovered 152 chimp killings, 58 of those they witnessed with their own eyes. In other words, a chimp researcher becomes aware of a chimp slaying another chimp more than once for every three years that he is in the field, and he will directly observe a chimp killing another chimp roughly once every eight years. This is in spite of the fact that it is difficult and risky to get close enough to the chimps to see what they are doing. Over time, the chimps' terrain becomes a marking ground for corpses.

This is typical of an array of primate species. In 1979 at a meeting in Bangalore, India, the Japanese researcher Yukimaru Sugiyama revealed to an audience of his fellow primatologists that he had uncovered a common practice among male langur monkeys. Young gangs, Sugiyama informed his listeners, "take over a harem of females by ousting the old leader and customarily killing the infants. They snatch them from their mothers' bellies, impaling them with their canine teeth."

All this eventually compelled scientists to ask a simple question: Might it be that the routine carnage observed among primates shows not that these species are like us but that they are quite different? By the early years of this century, the evidence for that had started to become unavoidable, and the field shifted once more. Anthropologists began to assert that humans were more *prosocial* than apes. This is a fancy way of saying that we are more inclined toward altruism, empathy, and collaboration. It was a belated admission that what they had been claiming during the 1970s was entirely backwards.

In 2019 this trend was highlighted by the publication of Harvard professor Richard Wrangham's book *The Goodness Paradox: The Strange Relationship Between Virtue and Violence in Human Evolution*. Wrangham did not go into all that much detail in describing the differences between chimp and human brains and what neuroscience has been revealing about each. But he did acknowledge the obvious fact that humans and chimps have been moving on very different evolutionary tracks. This was a great step forward.

Yet Wrangham's writings are carefully parsed. As he is a chief figure in the anthropological establishment, he seems not to want to engage in outright heresy. The sacrilege among their number is to suggest that humans are so unlike chimps and gorillas that trying to understand us through study of them is apt to lead us into a wholesale misunderstanding of human nature. Most specifically, this has led to grotesque errors by our politicians with regard to such matters as the risks of financial bubbles, the degree of transparency in international trade, sexuality and mating, and the likelihood of war with authoritarian nations. As we will see, it also leads to trouble in understanding such basic human actions as suicide, art-making, and even homosexuality.

That humans have a powerful inclination to promote ideas that are demonstrably untrue but generally accepted takes us back to Theophilus Painter. While we cannot deny his brilliance, he reminds us that our species is capable not merely of abundant wrongdoing but colossal folly. It's likely that Painter barely occupied himself thinking about the means by which he had obtained the genetic material for his slides, and his treatment of Sweatt is proof of the dangerous human capacity for either self-interested or cowardly obedience.

It seems that Painter was uncomfortable with his role as the public figure charged with keeping Sweatt out of the university that he was supposed to be running. Even so, he regularly offered up a steady diet of assurances to the segregationist board that had hired him, assuring them that Sweatt would never be admitted. Then, when the Supreme Court declared that segregation in colleges and universities was illegal, he abruptly shifted his position, displaying a sure-footed professionalism in telling reporters that, most certainly, the university would "observe the law." In the end, Painter did the right thing, if only because it was expedient.

Still, this is not the same as winning our mates by filling up our mouths with the warm blood of our neighbors' infants. Why do we reflexively—and accurately—describe such behavior as "inhuman"? How did we come to be almost incapable of acting that way? As we will see, humans have passed through two great waves of evolution. Through these our species separated itself from chimps and other primates. That's confirmed by a vast amount of new research, and, though a number of the views presented in this book are my own, above all else what you will find in this volume is a summary and an explanation of these revelatory findings. This is the work of neuro-scientists, biochemists, geneticists, experimental psychologists—and a few iconoclastic primatologists. Collectively they have provided a mass of evidence that conflicts with the notion that chimps and gorillas are all that similar in brain design, biochemistry, or behavior to people, along with hints about what this tells us about *our* nature.

Throughout this book, I am going to use the word "social" rather than "prosocial." This partly reflects my unease with jargon. It is prompted by another aim too, though. How and why we are different from apes and other primates—good and bad—is connected to our sociability.

I come to these questions from several directions. When I was a boy, I regularly listened to discussions of neuroscience at the dinner table. This is because my father, mother, and stepmother were psychologists and had been trained in it. That led my father toward the academic research for which he became known: laboratory experiments that examined the effects upon the brain of a wide range of drugs, including LSD, tricyclic antidepressants, and antianxiety medications. It drew my stepmother toward investigation of the effects of sex hormones upon male and female brain development. And it provided an underpinning for my mother's career as a therapist who treated the mentally ill. I have been a novelist and playwright, and their interests merged with my own when my career as a writer awakened my curiosity about how the emergence of spoken language shifted our evolution, altering our character and our consciousness.

One occasion crystallized my interest in the subject though. In a single moment I became aware of how wrong the primate model of human nature was. When it happened, I was already familiar with much of the research that is the focus of the next chapter of this book, studies showing how our brain is different from that of an ape. That wasn't what caught my attention. Rather, my interest was stirred by a pair of German Shepherds. The first was named Buster. His owner was a friend who made a small fortune in the apparel industry. Retiring in early middle age, my friend purchased a townhouse a few blocks away from the apartment where my wife and I live. Soon afterwards, he married and fathered a daughter. His life now revolves around his child and composing short stories for literary magazines. His other great love has been his dogs, Buster especially.

Because Buster suffered from bone cancer, my friend paid for his medical treatments. This included chemotherapy and the amputation of one of his legs. The loss of the limb did not deter my friend from walking him several times each day. No matter if it rained or snowed, he would take him out, and I sometimes ran into them in the harshest weather. One evening I spotted my friend taking Buster about as I was on my way to the grocery store. Returning home, I caught sight of another dog walker with an amputee canine, a chocolate beagle. That dog trundled forward on a contraption that permitted him to lay his haunches on a platform for support as he strode ahead, using his three legs.

Sometime later Buster died, and my friend adopted another shepherd, a rescue dog named Jasper. One afternoon I met up with them at a local park. While we chatted, my friend stroked Jasper, making affectionate remarks. Jasper listened intently. There is nothing terribly unusual in this, of course, and that is the point. It isn't only that dogs possess more talent for understanding our speech than chimps. They are interested in our thoughts and feelings. Their attachment to us reflects their evolution, which required them to become our partners as a means of survival. But our mutuality and our sense of kinship with them also arise from the parallel tracks in our evolution. Like them, we are domesticated animals and cooperative hunters. We also share these traits with dolphins and some species of whales, and we are probably even more like these cetaceans than we are canines.

This makes evolutionary sense. It would be *in*consistent with the theory of evolution if this were not so, and we were more like chimps and gorillas. Obviously, this conflicts with what we have been taught. We have been trained to believe that it is our status as primates that led us to our high intelligence and our capacity for civilized existence. From early ages we have been shown charts of human evolution. At the beginning—on the left of the picture—is a crouching, hirsute figure: an ape. He is the first in a line that literally rises up. Near the end of the chain of figures, we spy a prehistoric man in loincloth. Fully upright, he holds a spear. Farthest right is a modern human. These "ascent of man" pictures depict a precise series of steps leading to our present form with its implied perfection. However, evolution doesn't work like this. Dinosaurs led to birds, but they are also the ancestors of reptiles. The notion that humans should be like apes is based on a misunderstanding of evolution. Animals don't display similar traits or behaviors because they have common ancestry but because they have the same evolutionary needs.

Those who think of humans as primates are like people who drive past an airport, quickly turn their eyes to it, and conclude, based upon a cursory inspection, that the terminal's purpose is to serve as a giant parking lot. Just as the principal aim of an airport isn't to serve as a garage, though it may look like one, humans may somewhat resemble chimps, but we didn't evolve to be tree-dwelling herbivores. But this is what primates are. Simply put, the forces that propelled us to become humans represented a radical divergence away from a primate nature.

Our very different evolutionary demands required us to become a completely different animal. This is not a matter of conjecture. Over the last twenty years, new instruments like functional magnetic resonance imaging machines (fMRI) and CT (computerized tomography) scanners have allowed neuroscientists to examine the mind and observe its responses. This happened as geneticists were providing us with a grasp of its biochemistry. Their findings have overthrown most of the long-held assumptions of the primatologists. They show that our brain was devised to be unlike those of chimps and gorillas and that it shares many of its key features with those of other cooperative animals like dogs, elephants, and dolphins.

As I grew up listening to discussions of neuroscience, I have followed this revolution in the making, and this is my attempt to summarize this work for a general audience. But my desire to write about this was awakened by that afternoon in the park. I would even connect it to one moment—the instant before I parted from my friend. He leaned over and whispered in my ear that on the following morning he was driving to Pennsylvania to see his wife and daughter. As he had done before, he planned to leave Jasper with a housekeeper for a few days. He was careful that Jasper should not hear the word "car" as the news of his leaving would upset him, and Jasper knows what the word means.

As I walked home, I thought about my family. I was aware of the results of the experiment that demonstrated chimps have no inclination to act on behalf of other chimps when it requires nothing of them. My stepmother, now passed on, was exceptionally empathetic. So, as well, are my mother, my wife, and my cousin Max. Feeling for others drew my cousin and my mother toward work in hospitals. My wife displays it in her love of animals. This is not, however, a trait that is commonly observed among apes. Understanding the origins of these qualities and recognizing their implications—not all of which are good—is essential to grasping who we are.

The argument primate researchers most depend upon for their claim that we are primates is based upon a simple-minded view of our genome. When scientists first examined the DNA of humans and compared it with chimps, they reported that it was between 98 and 99 percent the same. Not every study has come to this conclusion though. That's because primatologists employ a somewhat debatable method of counting in order to come up with the generally cited figures of 1–2 percent. To arrive at the lower number, they throw out a range of genes known as indels. The reasoning used for ignoring the indels is that some of them are repeated strands of DNA whose role is not well understood.

This isn't all that is questionable in this line of argument. Each time our genes are copied and passed on there is a certain error rate in the copying. Thus, most of the mutations in our genome occur randomly, and a species' genes serve as a molecular clock. Alterations in the genome tells us how long ago two species parted from their common ancestors. While this clock is not as precise as tree rings are, its degree of variation is predictable. Several factors alter the movement of the hands. Among the most significant are the species' metabolic rate, its population size, and the length of time required for it to pass through each generation. This applies to the evolution of primates as much as other animals, and as we split from our common ancestor with chimps between six and eight million years ago and each human generation runs to roughly twenty-five years, the molecular clock tells us that we should be as genetically distinct from chimps as we, in fact, are. So, while primate researchers are fond of pointing out that our genome is more like a chimp's than a chimp's is like a gorilla's, they are merely referencing the clock and the fact that gorillas diverged from chimps and humans further back in time. It does not prove that we are more like them. The argument based on genome tells us nothing that the fossil record didn't already and little about our nature. Nor is this new information. It's something that scientists have known for almost sixty years.

To get a sense of how long ago our split was in evolutionary terms, consider that Soviet scientists seeking to tame silver foxes succeeded in domesticating them. In just 40 generations, they created a new species, a floppy-eared fox that behaved like an especially affectionate puppy. And our parting from chimps? That took place not 40 generations back, but more than 250,000 generations ago. Reflecting that and the advent of better technology, researchers have had to throw out the prior claims about our genetic kinship with chimps. New data from the National Institute of Health reveals that our genome varies from the chimp genome not by 1.4 percent but by 13.5 percent, and it's distinct from some apes by 27 percent.

In any case, let's go back to the matter of our genome and suppose that the variation from chimps really was of just 1–2 percent. Would this be a small difference? Most genes are guided by regulatory genes called promoter genes. Changes in even a handful of these produce radically different patterns of how proteins are synthesized and, as a result, in what sort of animal is spawned. That interested the researchers behind a 2007 study that found that many promoter genes showed evidence of differing from those of chimps not because of chance mutations but because of natural selection. The altered genes particularly turned up in the genes coding for our brains.

That brings up a subject of even greater importance: what part of the genome is shifting. The human brain is the part that has been changing the most, and one might say that it has a similar function to our regulatory genes. After all, just as they control a multitude of other genes, the brain controls the rest of us. Thus, substitutions in just a few of the genes that guide its design have altered our thinking and behavior in fundamental ways. That's been demonstrated through the Zoonomia Project, the largest comparative study to examine the human genome. It uncovered what appears to be the single most critical change in our DNA, the one essential puzzle piece explaining how and why we became modern humans. This didn't involve the loss or addition of even one gene. Rather, it was the deletion of just two base nucleotide pairs among the more than 100,000 within what's referred to as the LOXL2 gene. This small removal is what is known as a microindel or microdeletion. In other words, the most important evolutionary change—the critical switch in our design that led to our species' emergence—resulted from the most minute of indels. A change of just 0.0019 percent of the letters within the LOXL2 gene and of 0.000000067 percent of our total genome played the central role in transforming us into modern humans.

It is undeniable though that the greater number of our genes—for our muscles and bones, our skin and viscera—are the same as those of chimps. Thus, when we spend time watching them, we cannot but be struck by how much their expressions resemble ours. They *look* much more like us than species like whales, dolphins, and elephants. This is not only true of apes but also monkeys. This outward likeness inclines us to assume that they must be thinking and acting like us. But that they resemble us is not proof that they are like us, and if you get past the similarity in facial expressions, you have to acknowledge that we share much of our chemistry with all placental mammals. Yet even species that are so close genetically that they can mate and produce offspring frequently have entirely different social structures and patterns of behavior. That's the case with some of the best-known mammals, like tigers and lions, horses and donkeys, and wolves and dogs. These species pairs are remarkably similar genetically, and they can breed with one another. But the respective species are quite different. Male tigers hunt incessantly; male lions do not. Donkeys tend to be loners; horses are classic herd animals. Dogs are exceedingly tame; wolves are not. Two subspecies of zebras are herd animals, while among the third, the Grévy's zebra, the males are solitary and territorial. One could go on at great length with respect to this. The important point to be gleaned then is this: Given the marked differences between such very closely related species, people interested in them can only formulate reliable views by study of the species itself.

There are enormous differences between humans and chimps. These start with our anatomy. Chimps' arms are longer than their legs. They knuckle-walk. Their calf muscles attach directly to their heels. Their ankles can flex forty-five degrees straight back. Their feet can almost fold themselves over, compressing the two sides together so that they can fasten around tree branches when they are climbing. They have small buttocks. Their backs are absent of a lumbar curve. They do not possess tear ducts. They are two to three times stronger by weight, and their arms have up to five times our pulling strength. They are without our physical endurance. Their brains are between one-third and one-fourth as large. Where three-fourths of our cerebrum is constructed after we leave the womb, most of their comparatively modest cerebral cortex is fashioned before. And because chimp brains don't grow much after birth but their jaw expands dramatically, much of their head is made up of its lower half, and their forehead slopes forward. A chimp's skull winds up being stretched forward to meet his huge jawbone.

There are other basic differences. Many relate to reproduction and sexuality. One is shown most notably among bonobos. After bonobo females go into estrus, some will copulate 300 times, engaging in intercourse with all the male members of their troop. By contrast, human females don't go into estrus, and women can't be certain when they're ovulating.

Our appearance provides further clues about how different we are. On average men are 1.15 times as large as women. In the two chimp species, the ratio is 1.36 and 1.29, and in orangutans and lowland gorillas, it's 2.23 and 2.39. That's significant because the absence of differences in size—what is known as sexual dimorphism—tells us about our capacity for monogamy. Human babies also pass through a longer gestation period than chimp babies, and our offspring take more time to raise and instruct. In addition, like whales and dolphins, we were programmed to live to older ages, and human females are less likely to have twins.

These obvious differences in physical design, gestation, and lifespan are almost trivial though compared with those that emerge in how we act. Let's make a short inventory of some of the social patterns and behaviors that have been found among humans that have never been observed among apes. Though our list may seem obvious, I'm going to ask you to try not to erase it entirely from your mind as by the end of this book we will have seen how each one ties into the unique course of human evolution. More immediately, we'll see that disparate as they may at first seem, all of them have one thing in common.

1. *Ritual sacrifice of animals.* Throughout the world and across the length of human history, people have made offerings to the divine.
2. *Human sacrifice.* This was practiced by early civilizations on every inhabited continent. Among those sacrificing either children or adults were the Incas, Mayans, Mixtecs, Mongols, Egyptians, Chinese, Tibetans, Moabites, Hawaiians, Chichimecas, Aztecs, Natchez, Thuggees, Fon, Ashanti, and Celts.
3. *Suicide.* Self-destruction has been observed in other species. Most often this is through what is known as altruistic suicide. This is where a creature sacrifices itself to advance its kin. Altruistic suicide is regularly observed among insect species. There are also many recorded cases of suicide among cetaceans. As members of a group or even as individuals, whales and dolphins strand themselves.
4. *Pair-bonding.* Male chimps do not form romantic ties with females. Nor do they show interest in fatherhood— or even knowledge of whom their offspring might be. But pair-bonding has been observed in 226 mammalian species.

5. *Sleeping together.* Although chimp females construct tree nests in which they rear their young, males are not permitted in them. For chimps, sex is simply a procreative act.

6. *Division of labor.* Division of labor is one of the defining characteristics of "social" insects like ants, wasps, and bees. For example, in a honeybee hive, a female worker at different times may take on the tasks of removing waste, brood-nursing the young, making wax, guarding the honeycomb entrance, and foraging for food. Likewise, naked mole rats are separated into castes. In this way, the colony determines which rats transport soil, which care for the young, and which engage in tunnel construction.

7. *Naming of children and adults.* It turns out that humans are not the only species that names its children. Bottlenose dolphins name theirs. Some parrots also name their children, and a recent study employing machine learning indicates that elephants do too. But, as apes have no spoken language, they do not name their children.

8. *Blushing.* In Mark Twain's words, "Humans are the only animal that blushes—or needs to." Different explanations have been postulated for why people blush. The best seems to be that shame and shaming are fundamental human attributes.

9. *Crying.* We are the lone species that sheds tears.

10. *Polyandry.* Polyandry—the practice of a woman having two husbands—is rare. Nearly the only modern societies known to practice it exist in the Himalayas. There some women marry pairs of brothers. The purpose of the rite is not therefore to encourage promiscuity. However, it is a practice that is unimaginable among apes.

11. *Purposeful celibacy.* The Catholic Church has required celibacy of its priests since the First Lateran Council in

1123. Similarly, there are more than 100,000 Buddhist monks and nuns.

12. *Arranged marriages.*

To this list we might add an assortment of human activities that serve no obvious evolutionary function. Among these are the keeping of pets like tortoises and gerbils that do not act on our behalf, anorexia nervosa, Munchausen syndrome by proxy (the practice by which people poison those about them in order to attract sympathy), idol-making, temple-building, and burial of the dead.

You may now ask yourself: What connects these aspects of human existence? *All are social.* Think about crying. While people certainly will shed tears when alone, an audience seems to make the act easier, whether the observer is actual or imagined. This also applies to celibacy. While the position of incel may not be a choice, priestly abstinence is a conscious public gesture, one that is associated with other social roles and obligations.

A primatologist might respond by saying that chimps would engage in these acts if they possessed speech. Consider the reasoning though. This acknowledges a profound difference and then treats it as irrelevant because it's been acknowledged. What's more, many of these behaviors do not require the power of speech. Suicide doesn't. Crying, blushing, and anorexia do not. Nor does division of labor or arranged marriages. Even Munchausen syndrome by proxy doesn't require it.

Chimps have been known to make tools. Chimps have been seen honing tree branches in order to use them to pry out ants from dead logs. They grab rocks to smash the shell casings of nuts. They employ sticks to reach into the hollows of trees where bush babies hide. They can remember number sequences if they are rewarded with food for doing so. Some researchers even claim that they are better at this than people. But that they don't do any of those things on our list contradicts a basic claim continually made by primatologists. This is that apes are highly social. If this is so, then why don't chimps engage in these activities? And why are they so lacking in feeling for other chimps?

When the subject of empathy comes up, primate researchers are noticeably roundabout in presenting their ideas. De Waal, for instance, commences one of his discussions of chimp empathy with a lengthy account of a creature famed for its generous spirit who died long before the primatologist was born. The story includes no instance of the chimp sacrificing himself for anyone else, chimp or human. It merely describes him as warm and gregarious. Why did de Waal choose to write about this chimp that he never met? Undoubtedly, part of the reason is what you might expect: It's hard to come up with good demonstrations of chimp empathy. But there is a second purpose. That "chimp" was later revealed to be a bonobo, and in the last few years, de Waal and others in the field of primate anthropology have advanced a new line of argument for their claim that the study of chimps teaches us vast amounts about humans. Their current theory is that while *Pan troglodyte* chimps are vicious and egocentric, bonobos are not.

In putting forward this idea, de Waal was deviating from the general reading of his first book, *Chimpanzee Politics*. I say general reading as de Waal could have justifiably said that *Chimpanzee Politics* was misunderstood. Many people took it to mean that de Waal believed— underneath it all—that humans were almost exactly the same as chimps. In fact, *Chimpanzee Politics* points to a number of critical differences. For example, de Waal devoted several pages to a discussion of the way that females in the Arnhem Zoo are obligated to greet the troop's alpha male: by turning their back to him, bending over, and displaying their vaginas. De Waal was not suggesting that men and women interact like this.

Moreover, in his subsequent writings de Waal emphasized the kinship that humans bear to an array of non-primate animals. He also pointed to the derivation of the term "alpha male." First introduced in 1947 by Rudolph Schenkel in his paper "Expressions Studies on Wolves," the phrase did not originally refer to a physically dominant, swaggering figure like those alpha male chimps de Waal described in *Chimpanzee Politics*. Rather, it applied to whichever male happened to be the leader. Schenkel believed that this authority could come from many sources, physical aggression being but one. In addition, de Waal observed that the concept of pecking order originated with consideration of hierarchies among birds, not mammals. In other words, what de Waal was thought to be arguing in his first and most influential book and what he was actually proposing aren't the same.

Nonetheless, de Waal referred to these ideas in roughly their popular meaning in several subsequent books, even interpreting the outcome of the 2016 US presidential election in light of this conception of the role of the "alpha male." That reflected his continued belief in these concepts and the fact that a number of the book's central tenets have seeped into the culture. Through this, de Waal popularized the idea that there was a deep connection between the actions of the alpha male chimp and successful men. Both are presumed to be ruthless characters who obtain power and sex through their capacity for violence. That was depicted in *Chimpanzee Politics* in its extended account of the competition between the leading chimp "politicians" in the Arnhem Zoo. Their series of conflicts ended with the murder of a contender for power. Castrated, he had his fingers and toes bitten or ripped off. The story de Waal told was a frightening one, and he summed up his moral in blunt terms. "Among chimpanzees, hierarchy permeates everything." Power relations are "ingrained." "When political philosopher Thomas Hobbes postulated an irrepressible power drive, he was on target for both humans and apes." The viciousness and depravity and the harsh determination of position that de Waal saw among chimps served just as well as a depiction of human relations. If people were more subtle in their interactions, the basic grammar of social relations was the same.

This view is presented in support of the book's claim that humans share the chimp aversion to unselfish acts, and, in one form or another, these ideas have appeared in articles in general-circulation magazines and TV news programs. Through these they have been adopted by and influenced unscrupulous businessmen and ambitious politicians. Most famously, a Speaker of the US House of Representatives made a habit of recommending the book to his protégés. But, conscious that this literally Hobbesian view of human nature is in some measure wrong and surely unseemly, de Waal and his acolytes have recently been touting an alternate theory regarding the connection between men and apes. They have done this without rejecting their prior beliefs. It was, however, the spur to de Waal's account of that bonobo who lived in the 1940s.

Depending upon which authority you listen to, bonobos (*Pan paniscus*) are either a subspecies of *Pan*, the general classification for chimps, or a closely related species to the common chimpanzee, *Pan troglodyte*. The split between them goes back 1.5 to 2 million years. That was when the Congo River divided chimps north from south. As they are poor swimmers, the two sets of chimps moved onto different evolutionary paths. Bonobos, living in the forests south of the river, are sometimes referred to as dwarf or pygmy chimps, though they are not actually shorter. They are somewhat slimmer, or, in primatologist terminology, gracile. They have a different diet and social structure too. Much more than *Pan troglodyte* chimps, they tend to rely upon a diet of fruit, rather than leaves and insects, and they have a social structure in which females take a leading role.

De Waal referred to bonobos as the "make love, not war" chimps, and he put forward the idea that they are "gentle" creatures who can serve as another vision of human kinship with apes. They are "peaceful hippies." In a return of the old notion that humans are the bad and vicious hominids and not the cooperative one that builds hospitals for people and sanctuaries for chimps and gorillas, de Waal said that on occasion we are capable of being as "nice" as bonobos. This idea has permeated the culture. Thus, an Associated Press story that appeared in 2012, picking up from him almost word for word, asserted that bonobos were "kinder and gentler than chimps, and, well, us." It's a charming fable, and if you know nothing about the subject you might be taken in by it. However, field researchers who work with bonobos and neuroscientists who have examined their brains provide us with a different story.

It is a better one than that of the *Pan troglodyte* chimps. The field researchers report that bonobo camp grounds are not littered with corpses, and, as their society offers an enhanced role for female authority, female bonobos suffer far less at the hands of the males. Bonobos turn out to be better at reading social cues than other chimps, though they appear to be slightly worse at making tools. Their amygdalae are also somewhat more like human amygdalae. This suggests that they are more capable of empathy and compassion.

What else? In captivity, bonobos can be placed in a room that contains food *and* another bonobo, even one they don't know. In the language of primatologists, they are capable of co-feeding. This means that the larger bonobo won't immediately snatch away the meal or start beating up the smaller one to get hold of it. What's more, in the wild, bonobos will sometimes engage in mild forms of pair-bonding as females will occasionally sneak off with a favored male for extended periods of intercourse. This describes a more docile animal.

Yet the distinction is one of degree. Fighting of males with other males and females with other females is more common among bonobos than chimps. Bonobos "wake up ready for a fight." They are one hundred times as likely as people to engage in physical violence. You can get a sense for what their nature is by examining the best practices recommended for zoos where they are kept. Bonobos need to be separated from people by 5-meter-high fences (16.4 feet), and they are not under any circumstances to be given open access to roofs—even when they are in their own spaces apart from zoo visitors. Alternatively, a standard handbook recommends that around their dwelling areas water-filled moats be built. These are to be at least 23 feet wide and 5 feet deep. This is for a creature that cannot swim. Beyond this, secondary fences are recommended. These may be electrified. There are many more pages on the arrangement of secure areas in which bonobos can be placed when they are in conflict. These call for more fences along with steel locks for doors, iron mesh, and other impediments designed to keep them at an appropriate remove. By contrast, the fence height suggested for cheetahs is 2.5 meters (8 feet), and no moats are required. Those who toil with bonobos know that they are far more dangerous than big cats.

The research findings of neuroscientists align with this. The major areas of the bonobo's brain are not much different in size from those of chimps. The chimp cerebellum does appear to be a little larger than those in a bonobo. This would provide *Pan troglodyte* chimps with better coordination and balance. On average, chimps may also have a slightly larger hippocampus, though the data on this is not statistically significant. As one team of researchers who compared chimp and bonobo brains concluded, "The results suggest that the two species of *Pan* are quite similar neurologically." The investigators qualified this statement by saying that there might be "some" differences—though their data provided only modest support for this claim. In both species copulation is rare when females are not in heat—97 percent of chimp intercourse takes place when chimp females are near "maximal" swelling, and 95 percent of bonobo intercourse does. Moreover, in both the sex is unlike that of humans. Thus, it is reported that "as many as 50 copulation bouts with eight males during 1 day were observed [among *Pan troglodyte* chimps] and the swollen females copulated with up to eight adult males within several minutes. Bonobo females actively solicit sex from a range of males and may copulate multiple times per hour while swollen."

Nor are they any closer genetically to humans than *Pan troglodyte* chimps, and male bonobos are less inclined to form intergroup alliances with one another than *Pan troglodyte* chimp males. In the wild bonobos have never been observed engaging in what primate researchers refer to as "extractive foraging." Unlike chimps, they have no inclination to try and pry out ants from under rocks. Humans have to teach them this trick when they are in captivity.

Yet, in spite of all this, de Waal insisted that his research on bonobos indicates that *these* chimps especially demonstrate that the study of apes is instructive about humans. We are again faced by our problem of circular logic. In some respects, their behavior is more like people. It is still profoundly different. Let us look, for example, at the fact that bonobo females hold much more authority in social relations than *Pan troglodyte* females. Does that prove bonobos are like humans? Elephant society is matriarchal. Should we conclude then that people are most like elephants? If the test of a species' similarity to our own species is the level of female power, we might argue that black widow spiders are the most like people. Also, why should we ignore the first rule of evolutionary biology, which says that diet determines behavior? That tells us that bonobos are no more like humans than *Pan troglodyte* chimps and that neither is much like us. One wonders too: If bonobos are so sweet-natured, why do we know of at least one case when a mother consumed the remains of her five-year-old daughter?

This does not, of course, mean we cannot glean something from a study of bonobo behavior and of their evolution. In particular, primatologists have noted something distinctive about their environment that may be meaningful in understanding human ancestry: Unlike *Pan troglodyte* chimps, bonobos are not competing with gorillas for food. That's because no gorillas live south of the Congo River. As they needed less strength to fight off gorillas, that may have led them to a more slender build. It may be pertinent then that humans became slimmer and less violent when we evolved in ways that provided us with more and better supplies of things to eat. That appears to have been when we invented language.

The man who developed the generally used system of animal classification, Carl Linnaeus, focused on anatomical similarities. For this reason, he placed humans in a category he called Anthropomorpha. This order included monkeys. Employing his method of logic, Linnaeus claimed that the only "generic difference" between men and monkeys was in our faculty of speech. Modern taxonomists continue to use Linnaeus's system, and they draw their understanding from its focus upon common genetic ancestry and broad similarities in anatomy. The problem with this approach is that it systematically obscures and minimizes other connections among animals, which may be of greater significance. One of the most important is whether or not they are cooperative. Humans are. This is typical of meat-eating mammals that travel about in packs. Thus, dogs, wolves, hyenas, and many whales and dolphins are highly cooperative. So are a number of herbivorous herd animals like elephants and giraffes, as are many species of birds that flock together. But chimps, bonobos, gorillas, gibbons, and orangutans are not. In fact, within the entire order of primates there are only a few species that display much capacity or instinct toward working together. This has profound implications.

Without entirely letting go of that point, I'd like for a moment to turn our attention toward just one of these distinctive human behaviors I mentioned: ritual sacrifice of members of our own species. This is a cooperative behavior, if a very untoward one. Because it is seemingly such a curious behavior, it demands our special attention.

Charles Darwin was one of the most exacting researchers and theoreticians who ever lived. This was one of the reasons why he delayed presenting his theory of evolution for more than twenty years. He did not want to make it public until he had come up with explanations for every possible contradiction to his ideas. One issue that he spent an especially long time working to resolve was why hymenopteran social insects (e.g., ants, wasps and bees) were so cooperative and not competitive with one another. He eventually answered this when he came to understand that all the females of the colonies were closely related.

Darwin was also greatly interested in the matter of human evolution, and this is the subject of his highly speculative book *The Descent of Man*. Perhaps because a handful of its passages can be construed as racist it is rarely assigned in colleges and universities any longer. Yet it offers intriguing responses to many of the puzzles of human evolution. Nonetheless, suicide and human sacrifice are topics that he did not tackle. In fact, he actually implied in *The Origin of Species* that suicide could not happen, at one point stating, "Natural selection will never produce in a being anything injurious to itself." He had nothing whatever to say about human sacrifice.

Darwin's inability to grapple with these topics is easy to understand as they conflict with the basic tenets of evolutionary theory. Why would a creature choose to murder itself? Why would it kill its own young for no purpose but that custom dictated it? In addition, we must ask about a perhaps even stranger instance of this phenomenon than child sacrifice. We know of many documented cases of mass suicide. These have occurred from the time of Masada in ancient Israel to the Jonestown and Heaven's Gate cults in our own day. The kamikaze pilots of Imperial Japan serve as an example too. Such extreme behavior reflects a type of consciousness and psychology that is hard to square with the idea that humans are apes. Reports of dolphins and whales stranding themselves are commonplace. But there are no credible, documented accounts of suicide among chimps and gorillas. Why is this?

We know that mass suicides occur among members of cults. The term "social construct" tends to be thrown around in willy-nilly fashion nowadays. But cults are indeed social constructs, and the most notable thing about them is that they can persuade their members to engage in seemingly crazed behavior and to display blind obedience to the group and its leaders. Plainly, these acts and the beliefs motivating them are not in the victims' evolutionary self-interest. They do not help cult members produce more or better offspring. Why then do they do them? Or perhaps we should say, why do *we* do them?

The answer to these questions lies in our brains. The Zoonomia Project confirmed that this is where our genome is swiftly evolving. Necessarily then, it is the beginning of the path upon which we can grasp the nature of a vitally important animal that has been obscured from us: *Homo sapiens*.

CHAPTER 2

HOW EVOLUTION CHANGED THE HUMAN BRAIN

THE LAST TWENTY YEARS OF neuroscience research has provided us with overwhelming evidence that the human brain is different from the chimp brain, and increasingly the community of neuroscientists has come to the conclusion that persistent evolutionary pressures provided it with a number of essential design features that are either nonexistent or nearly so in primates. Yet primate researchers are willfully blind to all this. For the most part, they have gone on with their work as it was, pretending that none of these discoveries has happened, and the rest of us must now choose sides in an emergent debate.

The controversy is generational. The leading proponents of the idea that humans are best understood as primates are all over the age of seventy. Jane Goodall is ninety-one, Jared Diamond is eighty-seven, Richard Dawkins is eighty-four, and Richard Wrangham is a comparative youngster at seventy-seven. What these figures have in common is that they came to the study of anthropology and to consideration of the issue of humans and our relations to apes prior to the invention of the new technology revolutionizing the field. Thus, they developed the theories to which they remain wedded before the evidence behind our present understanding of the topic appeared. That revolution is particularly dependent upon three inventions: CT scanners, MRI machines, and DNA research.

Curiously enough, the first CT scanners were developed by EMI, the recording company that put out the Beatles records, and an argument can be made that John, Paul, George, and Ringo indirectly financed their creation. In 1974, Georgetown Medical Center became the first hospital to purchase one. Yet, while that machine offered detailed cross-sectional images of the body, it took eleven minutes for it to generate one image. Moreover, its pictures were rendered by a computer with just sixty-four kilobytes of memory. The result was a machine that could display soft tissue, including that of the brain, but which for more than twenty years after that was of little practical use to neuroscientists. Similarly, while the initial prototype of a full-scale magnetic resonance imaging (MRI) machine was developed and tested in England in 1978, the FDA didn't approve them for use on humans until 1984. Then there was the further problem that each cost millions of dollars. As such, *Forbes* has reported that "there were fewer than ten MRI scanners per million Americans in 1993," and there were even fewer machines than that in Europe and Japan. But their significance cannot be overstated. MRIs have allowed scientists to show the varying densities and the composition of different areas within the brain. This avenue for research was made of even greater value in 1990 when the fMRI (functional MRI) was invented. It allows scientists to see brain activity moment by moment. In this way, they can observe the levels of electrical stimulation, blood flow, and oxygenation as we are watching, reacting and processing information. They can see us thinking and responding to suggestions and to stimulation and record our responses as these take place.

That knowledge was augmented by new DNA technology. While scientists have known about DNA since the 1950s, the process by which it could be copied in a laboratory wasn't devised until 1985, and this technique, known as polymerase chain reaction, was not generally available to research scientists for nearly a decade after that. Employing it, geneticists can track genes over time, and they can search for connections between individual genes, behaviors, and the brain chemicals guiding these.

Altogether we know vastly more. Increasingly, the mind is not a black box, but a mirror. Whether the older generation appreciates this is another question. This appears to be the source for the hostility emanating from primatologists like Frans de Waal to neuroscientists like Harvard professor Steven Pinker. The latter figure inspired de Waal's "scorn" and a claim that Pinker is engaged in "serious cherry picking" in his analysis of human evolution and human nature. What was most likely at the root of de Waal's hatred is the arsenal of tools that Pinker and other neuroscientists have for investigation and the ways in which they are using these to question the presumption that humans are best understood as primates.

Pinker was born just seven years after de Waal, but he was trained in neuroscience. This means he is not only receptive to the technological revolution but also just young enough to have taken part. Indeed, Pinker heralded it at its outset, boasting about the new research being performed in *Time* magazine in 2000, "Using fMRI, a new scanning technique that measures blood flow, scientists can tell whether the owner of the brain is imagining a face or a place. They can knock out a gene and prevent a mouse from learning, or insert extra copies and make it learn better. They can see the shrunken wrinkles that let a murderer kill without conscience, and the overgrown folds that let an Einstein deduce the secrets of the universe."

That claim was recently proven in a most unusual way: A team of French-Canadian researchers used a series of scans of a woman's brain to reproduce the song she was listening to when the fMRI machine was charting her brain activity. (It was by Pink Floyd.) For primatologists, one of the most disturbing things about this is that it opens the brain up as something upon which the experimental method can be applied. That is a higher scientific standard for understanding than any that can be found through observations of chimps in the wild. Ironically, the eagerness with which primatologists strike at this sort of research can be read as an example of an impulse common among apes: territoriality.

Even so, we need not rely upon advanced tools to recognize how different our brain is from that of a chimp or a gorilla. There are many much simpler clues that we are more like certain species of dolphins—a creature without hands—than we are like apes. The most obvious is brain size. We can think of this in several ways. One is the amount of brain relative to overall mass. This provides an adjustment for the fact that larger creatures need more brain to control larger bodies. This ratio is referred to as encephalization quotient. By this measure humans rank first of all species and bottlenose dolphins second. A second means is simply to look at total size of the brain. Within the family of primates, this may be the most accurate method for predicting intellectual faculties. That suggests that use of encephalization quotient, which places humans first, overstates our preeminence, and dolphins and whales may be deserving of an enhanced appreciation for their intelligence.

There is a consistent pattern connected to this. All the mammals with the largest brains are herd creatures. This is true of whales, dolphins, porpoises, elephants, horses, hippopotami, and giraffes. Whales, dolphins, and humans also have the largest cerebral cortexes: the part of the brain involved in higher orders of consciousness and thought. There is one exception though. We are told that humans are not herd creatures. While we are obviously highly social creatures, we are said to have an ape's greater egocentricity and independence of mind.

But an abundance of clues tells us that our brains are arranged to take advantage of our capacity for speech and that this design calls for a creature with greater sociability. One pointer crops up through research on Asperger's syndrome. The afflicted lack skills at both language and socialization. Often the problem is that the focus of their mental activity has shifted from language skills to abstract and mathematical ones. In other words, the part of the brain most needed for language is also required for social cues, and, when these are not devoted to language as they are supposed to be, people struggle to read faces and to understand others' emotions. In many cases, Asperger's sufferers can perform exceedingly complex arithmetic computations, yet they have difficulty talking or grasping subtext. Simply put, our brains were engineered so we not only have the ability to talk but to interpret what others are thinking and feeling. The functions, it turns out, are bound up together.

Thus, as we diverged in our evolution from chimps, we moved onto the path of all other large-brained mammals with a highly developed cerebral cortex: We became much more socially aware. This is consistent with our observation that dogs can learn more words than chimps. A more social creature—in this case, a pack rather than a herd animal—has greater language skills than a more egocentric animal, a primate, though the latter is possessed of a larger brain. We have noticed this pattern in humans in other ways too. People who master languages and accents easily tend to be more aware of their peers' feelings and are more socially adept. They blend in better. Their voices sound more like their neighbors. They stand out less. Again: Socialization and language are intimately connected.

Further evidence for this can be found in examination of the brain's structures. Although chimp brains are much smaller, parts of their brains are comparable in size to areas in human brains. This includes the regions that control most of their sensory and motor functions. The differences show up elsewhere. One has to do with shape. Human brains are asymmetrical. That is, one hemisphere is bigger than the other. Chimp brains are not. In humans, the language areas, which are in the left hemisphere, are expanded. This is where the Broca's area is. It's the region that processes our wording so that it has syntax. We know that it has this function because when it is damaged, people can summon up words, but they come out in a sort of pidgin arrangement devoid of order. Although human brains on average are 3.6 times larger than chimp brains, the Broca's area is 6 times larger.

The Broca's area is one of several that deserves special attention. When neuroscientists began identifying the different regions in the brain, they thought of them in the way of memory chips in a computer's circuit board, and they imagined that each had a narrow, specific function. It turns out though that while each has a designated role that their effective operation depends upon their coordination. That's relevant in terms of another region of great importance in the development of human speech, a bundle of nerves known as the arcuate fasciculus. Brain researchers have learned that this highway links the Broca's area up with the other major language region, the Wernicke's area. Perhaps not surprisingly then, in humans the arcuate fasciculus is tremendously enlarged. And, as one study noted, "the projections of the human arcuate fasciculus reach beyond [the] Wernicke's area to a region of expanded association cortex in the middle and inferior temporal cortex that appears to be involved in processing word meaning." In chimps and other great ape species these routes are either much thinner or "not found" at all. This is to say that chimp and other ape brains seem to have been designed for shrieking and hollering but not for interpreting words or crafting sentences.

The researcher who has done the most to show the differences between human brains and chimp brains is a neuroimaging expert named James Rilling. A slim, elegant man in his forties with reddish brown hair parted on the side, Rilling serves as the head of the Laboratory for Darwinian Neuroscience at Emory University in Atlanta. One of the first ways in which Rilling and the others in Emory's neuroscience team drew attention was by showing how many more of the connections in the human brain were narrowly focused within an area called the inferior frontal gyrus. It is also involved in language processing.

A second discovery related to a part called the supramarginal gyrus. It plays a vital role in our capacity for empathy. So, as you might expect, among people with difficulty in recognizing social cues, its action is slowed. When scientists purposefully disrupted it, they noticed that their subjects struggled "to stop from projecting their own feelings and circumstances onto others." When the supramarginal gyrus wasn't functioning, their empathetic and intuitive faculties were set on mute. That has significant implications as studies reveal that in chimps these regions are smaller and less connected to other parts of their brains.

We can learn something as well from the position of the supramarginal gyrus. Located in the parietal cortex, one of its two halves is placed directly adjacent to the Wernicke's area. Like the Broca's region, this is in the brain's left hemisphere. The Wernicke's region is the area of the brain that takes the leading role in helping us to identify which words we wish to choose so that we can construct meaningful sentences. For this reason, the supramarginal gyrus's positioning is of great importance with regard to our ability to understand the feelings of others not as an inchoate sense of pain or pleasure but as a meaningful set of interpretations that we can sift through and deliberate upon. That's buttressed by its function over on the right side, where it plays a role in our ability to feel empathy.

The supramarginal gyrus connects not only left and right but up and down, and just below it is the limbic system. As the limbic system assists in our capacity to process feeling and emotion, the placement of these parts seems critical to what it is to be human. Through their combined action, we can sense others' feelings and then use the specificity of language to give this concrete form in the mind, something that apes, lacking language, cannot do.

Another area of the brain worth paying attention to is the cingulate gyrus. The term cingulate comes from a Latin word meaning "belt" or "collar" as it wraps around the corpus callosum. The latter section is the "tough body" connecting the two brain hemispheres along with the "reptile" part of the brain below it that generates our emotions.

In chimps and other apes, the cingulate gyrus is shorter and thinner. In humans and dolphins, it's a broader, more encompassing section. Its functions extend from helping to signal the heart to beat to regulating our emotions. When it's damaged, people become impulsive and are unable to gain a grip on their feelings. That we can do so, and that we are not bundles of uncontrolled rage, is tied to its healthy operation. That's demonstrated by a mountain of research. For example, it's been shown that the cingulate gyrus doesn't function properly in people suffering from Tourette's syndrome or among those who have obsessive-compulsive disorder and attention deficit hyperactivity disorder. In this way, we may have uncovered a clue to why dolphins, like people, are so good at reining in their feelings and why they can be rapt listeners. Further evidence that effective operation of the cingulate gyrus plays a pivotal role in human character is provided by researchers who performed brain scans on incarcerated psychopaths. It turns out that they have diminished activity in the back of the cingulate gyrus. More remarkably, whether it's working properly has been shown to be an indicator of whether an inmate is capable of avoiding the same mistakes; criminals with poor function in the rear of the cingulate gyrus are more prone to recidivism. Moreover, studies of women have revealed that there is a correlation between the size of the anterior cingulate gyrus and the capacity for empathy. It's also been found that this area is somewhat larger on average in women than it is in men, and we know that more men are psychopaths and that the size and function of the anterior cingulate gyrus correspond to this disparity. All this leads us to one conclusion: When people lack for empathy, they have brains in which the cingulate gyrus is like that of a chimp or gorilla.

This role of the anterior cingulate gyrus in guiding feelings of empathy and understanding isn't limited to humans. Neuroscientists had long been aware that the anterior cingulate gyrus is highly developed in rats. Intrigued by this, researchers at the Netherlands Institute for Neuroscience decided to do an experiment. First, they offered rats candy for pulling a lever in their cage. Then they shocked the foot of a rat in an adjoining cage when a rat asked for his treat. That caused the shocked rats to squeak in pain. The Dutch scientists noticed that some of the first group of rats, distressed by what they saw, stopped pulling the lever and asking for candy. However, when the researchers anesthetized the rats' anterior cingulate gyrus, preventing it from operating, *all* the rats went on requesting their sugar-sweetened snack. Without a functioning anterior cingulate gyrus, they no longer cared when they observed other rats in torment.

In a later chapter I'll discuss how human evolution and rat evolution are similar but unlike that of chimps and gorillas. What's significant is that in both rats and humans we know the anterior cingulate gyrus operates as an instrument for engaging our empathy. Yet it appears that it has an opposite role in chimpanzees. In human adolescents with poor impulse control, the gray matter of the anterior cingulate gyrus is less abundant. It's also been shown in adults that lack of gray matter in the region is associated with a likelihood of suffering from borderline personality disorder, post-traumatic stress disorder, and schizophrenia. Insufficient gray matter causes a breakdown in a person's ability to relate to others. Yet in chimps, it's been found that proportionately *more* gray matter in one hemisphere of the anterior cingulate gyrus is associated with *lack of ability* to pay attention to others. In effect, it appears that it has a function in regulating action and behavior in chimps, but it's been arranged for a different purpose to the one it has in humans and rats. Instead of arousing empathy and connection, it seems to have been constructed to orchestrate a chimp's indifference to those around him.

That might explain the discomfiture that de Waal and other primate researchers have displayed toward discussion of the anterior cingulate gyrus. After all, if this indicates what it appears to, then comparisons between chimps and the manner in which they conduct themselves and the way in which we conduct ourselves would be of little value—unless we were interested in the transactional dealings of voiceless mobsters in supermax prisons.

There is further confirmation of this in studies of a key region of the brain called the amygdala. Neuroscientists have long known that its proper functioning is critically important. The name comes from a Latin word meaning almond as each one is shaped like a small oval about the size of one. Sitting on each side at the base of our brain, the amygdalae coordinate our emotional reactions, especially those of fear and aggression. One of the most critical discoveries made by neuroscientists is that in psychopaths these don't have the connections with other parts of the brain that they ordinarily do and they don't function as they do in normal people. Thus, when you show most of us a picture of a horrible accident, the amygdalae will become excited as we experience a mixture of dread and pity. But when psychopaths are presented with the same images, the reaction in their amygdalae is lacking. Their amygdalae work to awaken feelings of rage rather than empathy. Moreover, a psychopath's amygdala—like his anterior cingulate gyrus—is often undersized.

Conscious of this, a group of neuroscientists at the University of California, San Diego, decided to do a study in which they looked at the amygdalae of humans, chimps, and bonobos to see whether there were differences. The team focused on the number of nerve cells designed for controlling the release of the neurotransmitter serotonin, a chemical whose levels in the brain determine our overall emotional balance. Perhaps not surprisingly, through postmortem inspection they found that in three of four regions human amygdalae had twice as many nerve cells designed for controlling the amount of serotonin secreted as there were in the chimps. (There were also dramatic, if somewhat smaller, differences in the number of such nerve cells in bonobos.) The team's conclusion: Absent of these, chimps have been programmed to respond to day-to-day situations with fear and violence, and, great as the differences between us and them are, this will be evident in the behavior shown by humans and chimps from earliest childhood.

These are hardly the only parts of the brain that show how our evolution has produced a quite different nature from those of chimps, bonobos, and gorillas. There are at least three more regions that must be discussed. The first is the caudate nucleus. It's buried deep within our skulls. In the same manner that the cingulate gyrus wraps around the corpus callosum, the caudate nucleus forms a girdle enveloping the thalamus. The thalamus is the "master gland," a sort of chief executive officer for the endocrine system. Directing and controlling the movement of all the hormones produced by the body, the thalamus also helps to guide sensory and motor responses. Hence, the primary role of the caudate nucleus is to provide final instructions and oversight for the body's master gland. In this way, the caudate nucleus helps to coordinate an enormous number of our functions and feelings. These range from physical tasks to our powers of speech, our sensation of hunger, and our emotions of love and affection.

In different ways, this attracted the interest of another group of neuroscientists. Most relevant to us are a research team at Kent State University in Ohio and a second group at Emory University. The Kent State neuroscientists wanted to see how the actions of the caudate nucleus were connected with working versions of the FOXP2 gene—a gene that's needed to produce human language. To that end, the Kent State team examined brain cells from the caudate nucleus in humans with and without functioning FOXP2 genes, along with mice and six species of primates, including chimps. Their particular concern was whether the action of the FOXP2 gene allows the nucleus to stop the secretion of a critical neurotransmitter. It turns out that it can and that this is a "species-typical characteristic" of humans, one that no primate possesses. In its absence it seems that speech commands are hyperactive: They go haywire. In follow-up research, the Kent State neuroscientists found that the biochemical structure of the caudate nucleus was entirely different in human brains from that found in primates. In their view, humans have a command-and-control structure for feelings and emotions that is unique.

The Emory team uncovered something else. The head of the group, Gregory Berns, is a dog lover, and he was curious to know why dogs are so capable of showering people with love. We'll discuss his research at greater length in another chapter. For now, we should observe what Berns and his fellow neuroscientists discovered when they asked dogs to respond to different tasks using word commands. To make sure that the responses were real, they threw in some invented words the dogs didn't know. That had little effect. But the words they did know caused a spike in activity in the caudate nucleus and the thalamus. Since the caudate nucleus guides parts of language processing in humans *and* many of our emotional responses, including our feelings of romantic love, this seems not to be an irrelevant point. That seems especially notable as the caudate nucleus has an enhanced role in dogs, and we know that chimps and gorillas don't pair-bond.

Granted that dogs are rich in love and affection, that words spoken by an adored figure have the power to activate a dog's caudate nucleus may not be a coincidence.

Neuroscientists have also shown that a part of the brain known as the dorsal medial prefrontal cortex assists in our awareness and consideration for those we don't know. And guess what? Although brain-imaging tests showed many similarities in the way a related region called the precuneus functions in chimps and humans, they noticed that in the chimp brain the dorsal medial prefrontal cortex was markedly less active. It seems that if someone is a stranger, chimps have been wired not to think much about them.

One last region of special note is a part of the brain called Brodmann area 10. Seated within a less easily accessed portion of the prefrontal cortex, it plays a central role in our capacity for planning, paying attention, speaking, listening, and showing self-awareness. In apes, it's smaller and less developed. That reflects a general trend. One especially expanded part of the human brain is the prefrontal cortex. Indeed, in humans it's more than six times larger than it is in chimps, and in both humans and cetaceans these parts of the brain have not only more gray matter but more white matter. The white matter helps with forming the connections among the different areas. In humans, it's been found that damage to the prefrontal cortex leads to loss of inhibitions and the absence of self-control. That's so true that it was one of the first things that psychologists ever discerned about the functioning of the human brain. They learned this through a famous accident. In 1848, an iron tamping rod splintered the prefrontal cortex of a railway construction foreman named Phineas Gage. In the immediate aftermath, parts of Gage's prefrontal cortex were visibly leaking out of his skull. Yet subsequent investigation showed that no other part of his brain had been damaged, and after less than a year he was well enough that he applied to return to his old job. He could still walk, talk, and perform all the other basic functions of everyday life. But his employers found that they had to let him go because he had suffered a dramatic change in his personality and nature. Without a properly working prefrontal cortex, he had lost his social acuity and his ability to regulate his moods and control his behavior. He was no longer able to plan out most of his actions, and he had become irascible, unpredictable, profane, and indifferent to others. Psychologists soon came to understand: The development of a large and active prefrontal cortex plays an integral role in what it is to be a human.

This is part of a general pattern, affecting other vital regions of the brain. The human mind has been changing rapidly, and the more it changes, the less it is like a chimp's. That was demonstrated by a study undertaken by a team at the University of California, Davis. They sought to determine what part of the human genome was shifting fastest. In total, they identified 202 areas in the human genome that were rapidly evolving. One stood out above the others though: a string of DNA that they named Human Accelerated Region 1, or HAR1. The researchers looked to see what it was responsible for. It turned out that it in combination with a second region known as HAR1F it coded for the development of the cerebral cortex. For much of the animal kingdom, HAR1 seems to be a standard, nearly unchanging strip of DNA. Indeed, for chimps and chickens 116 of the 118 letters in the chain are the same. However, in humans 18 of the letters have been flipped. Reflecting our greater intellectual needs, it's been transformed. Connected to it is still another crucial section of DNA. This is known HAR1A, and neuroscientists found that it guides the development of the fetal brain. It also appears to assist in its elaborate folding. Likewise, it's been discovered that the human cingulate gyrus relies upon the action of a chemical called calretinin, which acts as a stop button for when its excitatory nerves are overstimulated. Calretinin exists in much lower concentrations in the cingulate gyrus of chimps, gorillas, and orangutans, and it is not present at all in the cingulate gyrus of other primates. But what may be most important of all are those two absent nucleotide pairs identified by the Zoonomia Project. In the vast majority of placental mammals, they are located on the LOXL2 gene. In all, the gene has 107,224 base pairs. Yet the removal of just those two base pairs alters the code. The result is that our brains expand throughout childhood, adolescence, and young adulthood, while chimp and gorilla brains don't continue growing past the first years of life.

Perhaps even more remarkably, the differences between the brains of humans and other apes go beyond patterns of development, overall size, and chemical composition. They extend to cell design. That's seen in two types of specialized neurons, which can be found in multiple regions of the human brain. These seem to be critically important to its function, and each offers additional hints at why we are so unlike chimps and gorillas.

One of these leads us toward a fundamental question in neuroscience: How is it that humans appreciate that other people and animals, as separate beings, are thinking their own thoughts? This is referred to as the theory of mind problem. In recent years, neuroscientists have been working to resolve the issue through their study of a crucial type of cell called a mirror neuron. These are turned on when we observe another person performing an action or displaying a facial movement. An odd effect of this is that when we see someone hit a tennis ball mirror neurons controlling our hand and our arm light up as we seem to automatically and sympathetically process the observation of the gesture. Our brain teems with these mirror neurons, and they have been identified in many sections of the brains of songbirds, cetaceans, and monkeys. Most neuroscientists believe that these neurons play a role in the ability to empathize, and a number of studies support the idea. That mirror neurons crowd up in the anterior cingulate gyrus offers evidence for this. Yet mirror neurons have been identified in only about half as many brain areas in chimps as in humans. It's also intriguing that mirror neurons can be found in songbirds. After all, for a song to persuade another bird to mate it must be understood as a message. If it's perceived as the auditory equivalent of a sunset—a pretty aspect of the surrounding world that has no intentionality behind it—then it will not work. A female bird must realize what the singing male is trying to tell her and why he is expending the effort. This may be why mirror neurons have been found in humpback whales as males of the species also woo their mates by singing. In each case, this requires a theory of mind. That's further indicated as there are mirror neurons in the anterior cingulate gyrus of rats, and they activated themselves in that experiment in which the rats displayed concern for the other rats who were shocked.

In addition, neuroscientists have discovered that the anterior cingulate gyrus has a large number of another type of highly specialized brain cells. These are called von Economo neurons (VENs), or spindle regions. These oversized cells assist with the speedy movement of chunks of information. Named after the early twentieth-century Austrian brain researcher who first identified them, VENs are abundant in human, cetacean, and elephant brains. They are far fewer in the minds of chimps and gorillas. These cells work within the anterior cingulate gyrus to pass along messages to Brodmann area 9, along with area 10. Area 9 helps us deal with "cognitive dissonance." In conjunction with area 10, the two regions operate to force us to admit our errors, ponder our past mistakes, and map out better plans for the future.

Chimps and gorillas are generally lacking in VENs in other parts of the brain too. In fact, one comparative study of the fronto-insular cortex—which plays a role in constructing our consciousness, our self-awareness, and our ability to relate to others and to feel pain—revealed that an average adult human had 82,855 VEN cells within it, an adult gorilla had 16,710, an adult bonobo had 2,159, and an adult *Pan troglodyte* chimpanzee had just 1,808. The typical primate brain, it seems, is poorly developed for these tasks. In the case of elephants, the VENs appear to be particularly developed in the anterior cingulate gyrus, the region that as we know plays a central role in processing feelings of empathy.

There are additional reasons to suspect that the relative absence of significant numbers of von Economo neurons in apes greatly affects their ability to grasp subtext and to understand the thoughts and feelings of their kin. In the last few years, neuroscientists have begun to realize that problems with the distribution and function of the spindle regions are strongly predictive of autism. That prompted neuroscientists to conclude that they play a central role in our ability to make rapid, intuitive judgments in social situations. Thus, animals with masses of von Economo neurons appear to have been prepared for a wholly different evolutionary track than those without them.

Taken as a whole, what we seem to be seeing is convergent evolution. This phenomenon is a bit like what happens when two groups of scientists independently solve an engineering problem in the same way. Nature also works like this. A striking instance is shown among vultures. Though Old World and New World vultures are nearly indistinguishable, they are unrelated. Old World vultures are cousins of eagles, while New World vultures are descended from storks. Yet because they hold the same evolutionary niche as high-flying scavengers, they evolved to look and behave alike. Similarly, the brains of the animals that needed to be the most communicative, intelligent, and social developed parallel structures. The result was that the parts that provide the capacity for ordered emotions, understanding of others, and empathy are less developed, less capable, and smaller in chimps, bonobos, and gorillas, and more refined, bigger, and more efficient in humans, cetaceans, and elephants. One study of the human genome that compared it with those of dolphins found that we had undergone a parallel pattern of development with respect to genes for overall mental development, sleep functions, metabolism, and even our mental disorders.

But why, you may ask, is this so important? To answer that question, let's examine one of the great problems of our species: war.

CHAPTER 3

WHY WE FIGHT WARS

EVOLUTION NEEDN'T BE RAPID. BUT it is a relentless instrument for molding creatures to fit the demands nature places upon them, and it can cause animals to take on traits and modes of living that are wholly unlike those of the other creatures within its order. Take marsupials. We tend to think of them as cuddly plant-eaters. But 35,000 years ago Australia had a species known as *Thylacoleo carnifex*. An apex predator, *Thylacoleo carnifex* filled the same evolutionary niche that the big cats do in Africa and Asia, and it's sometimes referred to as the Australian lion. Convergent evolution provided it with a huge jaw, sharp teeth, big claws, and tremendous quickness. As you can see from a picture, it looks like a lion. Yet it was a marsupial. The females had a pouch in which they carried their young. This does not mean that they were much like koalas. They did not sleep in trees. They did not eat leaves. Nor did they hop about like kangaroos.

In the same way, there once existed a huge big "cat" in South America known as *Thylacosmilus*. Though it resembled the saber-toothed tigers of the Old World, they were unrelated. Like *Thylacoleo carnifex*, it was a marsupial, and the mothers carried their young in small pouches below their waists.

One could cite thousands of such cases in which an animal came from one order but took on the characteristics of another in response to evolutionary pressures. These forces are all around us, and, if we bother to examine them, we can figure out how they operate.

Carnifex thylacoleo, an extinct marsupial (Australia).

Thylacosmilus (South America).

Saber-toothed tiger (Europe and Asia).

Consider one of the subjects that I introduced in the previous chapter: the differences between horses and donkeys. Horses evolved from an equine species that lived on the Russian steppes. Donkeys descend from a wild ass found in the arid regions of North Africa. Because fleet herbivores living in open grasslands have plentiful food supplies, they are multitudinous. So horses must herd together to avoid the predators that surround them. Equine herbivores living on the margins of deserts are relatively few, and they will tend to split up in order to capture the limited food supplies available. Consequently, donkeys tend to be loners. The same pattern is observable among zebras. The famously lonesome Grévy's zebras live in near-desert regions of East Africa. The other two subspecies of zebras trek through verdant savannahs. Is it any surprise then that horses and common zebras are herd animals, and donkeys and Grévy's zebras are solitary and territorial? To assume that people would be like chimps and gorillas when our evolutionary requirements have been so different is peculiar. It's like thinking that since Rolls-Royce started out as a car company that the jet engines it manufactures for planes must have pistons.

Our thirty-two teeth are like the pouch of the marsupial lions or the Rolls-Royce nameplate. It is an incidental inheritance, neither a demonstration nor a proof that we are a third species of chimpanzee, as one widely read evolutionary biologist suggested. Another of the incidental inheritances we walk about with are the genes that control facial expression. These make us think that chimps and gorillas are like us in their cognition and their actions, but the evidence provided by neuroscientists and biochemists in the first case and primatologists in the second shows that this is not so. Having spent time observing monkeys in Central America, I understand the impulse to think of primates as alike to us because they resemble us. But this is hardly proof that they see the world or behave as we do.

There are critical lessons here. One relates to the most worrisome of all human activities: war. Primatologists have convinced our intellectual elites that warfare is a natural outgrowth of our primate impulses for violence and aggression. Hence, we are apt to think that it is inevitable, and we are told that there are only a few means to avoid it. One is a commitment to diplomacy. This is to be aided through the creation of international organizations, like the United Nations, run by wise men dedicated to resolving disputes between the countries of the world. We are also led to believe that we can reduce the dangers of fighting by facilitating trade, which will create a degree of mutual dependence. These are said to be tools that will help us bring a measure of civility, dispassion, and calm to situations that otherwise would end in bloodshed. The persistent failure to prevent war shown by organizations like the UN is said to be a demonstration of how strong and innate our impulse to fight wars is.

The "proof" that chimp troops attack one another in the way that human armies do is an event known as the Gombe Chimpanzee War. There are between 170,000 and 300,000 chimps in the wild, and researchers have found that the more densely populated a chimp terrain is, the more murderous they are. Let's keep those facts in mind when we consider the chimp "war" that took place between 1974 and 1978. The conflict began when a troop of chimps split up, dividing into two new troops. One consisted of nine adult chimps and their offspring. The other was composed of twenty grown chimps and theirs. When a fight ensued over the few square miles of territory in which they lived, the male chimps in the smaller group were gradually picked off.

This is consistent with the high rate of chimps killing one another when they are in close proximity. War is something different. It is a brutal contest involving organized slaughter on a mass scale, and while humans are not the routinely murderous animal that chimps are, we do methodically construct immense armies that engage in wholesale killing. Yet during the entirety of the Gombe Stream War, there were just eleven deaths. These took place over four years. That means that two groups of supremely violent animals in regular contact and continual conflict with one another averaged fewer than three killings each year. This was a gang fight, not a war. And that is to be expected. After all, for all the chatter of ape competence and intelligence, they cannot be trained to perform intricate operations requiring planning and collaboration.

We can get a better sense of what species are capable of war through examination of the ones that assist our military. Among these are the marine mammals employed by our navies. During the Iraq War, the US made use of a squadron of sea lions trained to guard our ships, and, in its struggle with Ukraine, Russia is using dolphins to protect its Black Sea fleet from underwater divers. There have also been repeated instances in which killer whales have engaged in coordinated attacks on boats. In striking at us, they show that they will not only make war against each other but that they can even organize campaigns against us. At the same time, we happily and safely play with this same animal at ocean parks, and we can train them to perform tricks for us in return for treats.

This is not a contradiction, and it gets us to a larger point: We are susceptible toward the impulse to war not so much because we are innately violent in the manner of primates but because we are tame and cooperative. Our empathy and altruism go hand in hand with our obedient nature and our ability to work together. So, it should be no surprise that the animals we have involved in our conflicts—not only sea lions and dolphins but horses, camels, pigeons, and elephants—are docile and cooperative. Only tractable animals that work together, like humans, are capable of fighting wars.

For further proof of this think for a moment about the species that is most notorious for engaging in wars of annihilation. Which one is that? Ants. In the words of the famed entomologist E. O. Wilson, "Ants are the most warlike of all animals, with colony pitted most violently against colony of the same species. Extermination is the goal for most, and as a rule larger colonies defeat smaller ones. Their clashes dwarf Waterloo and Gettysburg." This makes sense as ants are obedient, collaborative, and social to a degree that far exceeds humans, and war is an activity of biddable and cooperative creatures.

A much talked-about experiment provides further evidence that our obedience lies at the core of what turns us into "good Germans." Many readers are likely familiar with the famous—or notorious—research of Stanley Milgram. In 1961, Milgram recruited test subjects and asked them to give electrical shocks as part of what was supposed to be commonplace psychology research. Although there were no actual shocks being sent and only actors faking the pain that the electrical charges were said to inflict, Milgram's study revealed that ordinary people could be persuaded to participate in acts of torture if they believed that what they were doing was justified by science and if they had the encouragement of an authority figure.

First performed on the Yale campus in New Haven, Connecticut, the study asked volunteers, paid a minimal sum, to administer the supposed shocks in order to find out whether doing so improved recall. The volunteers were asked to begin the experiment by giving another person a 15-volt charge. Then they were instructed to give successively higher levels. The amounts increased by 15-volt intervals. With each boost, the person to whom the shock was given displayed increasing distress and pain. At 150 volts, the actor yelled out; 82.5 percent of the volunteers reached that level. From that point on, the actor receiving the fake electrical charges displayed more and more grotesque and horrible reactions. Finally, at 330 volts, the actor stopped hollering. Instead, he pretended that he might have been knocked out by the anguish of what he had been subjected to. Nevertheless, on average, almost two-thirds of the volunteers raised the level up to what appeared to be the highest level: 450 volts. This was where the test equipment was marked with a skull and cross-bones and XXX—extremely dangerous. The study's results have been duplicated, and researchers have found similar results among volunteers of different backgrounds, sexes, and ethnicities in countries around the world. When an authority figure assures people that they should do something inhumane, large numbers of people show their natural obedience and do as they are told.

Yet one of the most striking findings of the Milgram experiments garners relatively little attention. When Milgram asked the volunteers whom he brought in to form into groups of three and direct the research themselves, not one person went ahead and placed the high-voltage shocks on the actor playing the part of the test subject. It was only when a guiding authority figure instructed the volunteers to engage in torture that people did so.

Because so many primatologists see humans as a closely kindred species to chimps, they mistakenly think of war-making as a general human tendency and not a specific consequence of forms of government that empower the worst among us and then permit these men to employ our herd instincts for their own ends. If primatologists were right that our wars derive from the same impulses toward dominance and aggression we see in chimps and gorillas, then the Milgram volunteers acting alone would have behaved as the first study participants did. To the same extent, if war were a natural and inevitable outgrowth of a primate's desire for overlordship, then chimps and gorillas would methodically compose themselves into primate legions and slay each other in mass eruptions of violence alike to our wars.

Altogether, this makes little sense as an explanation for the behavior of humans or primates.

That humans need not be engaged in constant warfare is easy enough to show. Think for a moment about two of the longest borders in the world: the 5,525-mile dividing line between the United States and Canada and the 2,607-mile perimeter separating Russia and China. The first is undefended. In the second case, Russia and China claim to be friendly neighbors and allies, and in recent years they have been conducting joint military exercises. China has also been supplying Russia with weapons to use in its fight with Ukraine, while Russia has been selling China discounted oil. Their seeming friendliness is a recent phenomenon though. At one point Russia had thirty-four divisions stationed along their frontier, and China had fifty-nine, and their border continues to be guarded by hundreds of thousands of troops. More tellingly, in March 1969 the Soviet Union initiated plans to use nuclear weapons against China, and it was only the secret intervention of the United States that stopped this from happening.

Yet the situation of the US and Canada is hardly unique. European nations that went to war with one another for hundreds of years no longer bother to patrol their frontiers, and no one seriously expects Germany and France, or Sweden and Denmark, to fight. Once women were given the vote, democratic nations stopped going to war with one another. This type of government prevents authority figures from leading their people into battle. Hence, not a single war has occurred between these nations. Nor have any even come close to it. By contrast, wars still regularly break out in the Middle East, Africa, and Asia—the regions where dictators rule. War can then be defined as a process by which the destructive impulses of ordinarily tame and obedient young men are harnessed and directed by cruel, avaricious males. Dictatorship facilitates this.

In the final chapter of this book, we will look at this question in more detail. What's critical to bear in mind is that just as marsupial lions never were deprived of their pouch, we have held on to some aspects of a primate appearance, but we have assumed a different evolutionary niche and different instincts from those of the animals we are most related to. To understand why that is—and to gain a proper understanding of our propensity for committing suicide, our love of fashion and sports, the phenomenon of preferential homosexuality, and so many other deeply embedded human activities and instincts—we must first arrive at an understanding of the path by which, like those extinct marsupials, we changed from herbivores to meat-eaters, ultimately emerging as apex predators.

CHAPTER 4

HOW WE BARELY SURVIVED AS HUNTERS, OR WHAT MANATEES TELL US ABOUT US

THE DISCOVERY OF EVOLUTION WAS directly connected to the way in which diet influences chance mutations. Finches have been called the "emblems of evolution." That's because Charles Darwin's interest in the subject was awakened by the variations in the size of the beaks he saw on the finches in the Galapagos Islands. Those were caused by differences in what they eat, and we now know that differences in intelligence among a broad range of species strongly correlates with dining habits too. That tells us, of course, that we are unlikely to gain much understanding about humans by studying primates.

Yet within the last thirty years, primatologists drew the opposite conclusion, and this has been reflected in their language. Up through the 1980s, the term "hominid" was used to refer to present-day humans and either our direct descendants or our recent rivals: Neanderthals, Denisovans, and, going further back, *Homo erectus* and *Australopithecus*. Then, reflecting the claims of authors like Jane Goodall and Frans de Waal, as well as a general liberality in thought, it became common among primatologists to call all great apes hominids. The suggestion is that all are like people. Yet, except for chimps, they are not even very close to us in the chronology of their evolution. Humans diverged from a common ancestor we share with orangutans between fourteen million and eighteen million years ago. That is not only a long time to wait for a bus but also a great passage in evolution.

Even so, if we are to try to learn about people through the study of primates, it is not obvious that we should begin our investigation with chimps or gorillas. In many respects, after all, we are more like orangutans. The old Clint Eastwood movie *Every Which Way but Loose* concerned a truck driver who rode around with a pet orangutan. That's highly implausible but not ridiculous. The notion that one could share the cab of an eighteen-wheeler with an adult chimp or gorilla *is* ridiculous.

To get some further sense of how unlike chimps and gorillas we are, think for a moment about the differences between housecats and lions. The evolution of *Felis catus* diverged from that of *Panthera leo* more recently than humans split from our common ancestor with chimps. It was about five million years ago. But in one obvious way the smaller of the two housecats my wife and I live with is less like a lioness than we are kindred with chimps. At five pounds, she is one-sixtieth a lioness's weight. Chimps are of similar size to humans. However, in behavior and appearance big cats and housecats are much more alike. While it is illegal and inadvisable to have a pet tiger in your backyard, it is not only possible but not even that unusual. It has been plausibly suggested that more tigers are residing in private homes and facilities in the United States than in the wild. By contrast, as we have already mentioned, no one can safely live with an adult chimp or a gorilla.

The similarities among cat species are great. All race around on four legs, and their basic engineering and means of locomotion are the same. Thus, they have similar coats and eyes, and their ears, skulls, and noses are shaped the same way. What else? They have excellent sense of smell and sharp if primarily monochromatic vision. They love to sleep during the day. They stalk their prey. They have claws. They excel at climbing and sprinting and possess superior balance, jumping ability, and agility. They lack for endurance. Their dentition is similar as is their carnivorous diet. The four *Panthera* species (lions, tigers, jaguars, and leopards) roar. Both tigers and jaguars can swim. In fact, although we identify them as separate species, so closely related are they that they can interbreed. The smaller New World felids—cougars, ocelots, and lynx—can interbreed as well, and housecats and ocelots look so much alike that people have mistakenly taken ocelots in as rescue cats only to realize to their displeasure that they were not so docile.

Even so, no such likeness exists between the other surviving "hominids" and *Homo sapiens*. No great ape can produce live offspring with any outside its genus. Only gorillas can mate with gorillas, orangutans with orangutans, and so forth. There are horse and zebra crosses called zorses, tiger and lion crosses called ligers, and grizzly bear and polar bear crosses called pizzlies. There are even hybrids of beluga whales and narwhales: narlugas. But because we have two fewer chromosomes than chimps and six to eight million years of different evolution, we cannot interbreed with chimps—as they cannot cross with gorillas or orangutans.

We have already discussed some of the obvious differences in our bodies and those of chimps. Let us now consider the ways in which the faces are distinct. The most obvious difference appears with the human nose. It is not shallow and recessed, but prominently set forward. One cause for this is our faculty of speech. That led to the expansion of our pharynx, which required a much larger nasal cavity, one placed farther ahead. There are also obvious differences in our ears. Because chimps evolved to live in tropical forests, they can hear higher-pitched sounds than humans—the noises of other creatures. Conversely, we have large eyes, and it has been shown that even when we are babies that we spend much more time making eye contact with adults than chimp babies do with chimp adults. In fact, studies indicate that from very early ages apes have a natural impulse to ignore the expression in the eyes so that they can better track the possibly threatening gestures coming from above or below other creatures' lines of sight. Humans, on the other hand, have a natural inclination to focus our gaze upon each other's eyes. This helps us, of course, to know what our peers are thinking and feeling.

To assist with this, the sclera of our eyes is white. The opposite is true of chimps and gorillas: Only a small flash of the white can be seen at the rims of their eyes. Most anthropologists believe this is related to the fact that they are less cooperative. Proof that the white in our eyes serves to assist us in working together comes from the other animals with a visible white sclera: some dogs, most coyotes, and many cetaceans—animals that hunt together. Like humans, they must be able to see in what direction their peers are gazing and what they might be thinking. Since chimps and gorillas rarely work collaboratively, and they are more often engaged in egocentric and deceptive behavior, it is not advantageous for them to have eyes that can easily be "read," though it is a necessity for us. In this sense, the fact that we have what are sometimes called "whale eyes" is alike to the human propensity for blushing. It serves our interest that we expose our feelings and our intentions as we are social, cooperative, and domesticated.

However, none of this tells us the actual record of these changes: *when* they occurred. To understand that we must make a brief survey of the lengthy period of our prehistory. As we don't have devices like fMRI machines to test what happened, this is speculative in a way that neuroscience is not. Yet the fossil record and some hints from genetics provide us with vital clues, and what these tell us above all else is that the ancestors of modern humans passed through two periods of sudden and dramatic change. You can see this on a chart of the weights and brain sizes of different hominid subspecies.[1]

Australopithecus	85 pounds	440 cubic centimeters (2–4 million years ago)
Homo habilis	75 pounds	640 cubic centimeters (1.65–2.3 million years ago)
Homo erectus (Java man)	127 pounds	930 cubic centimeters (600,000–1.7 million years ago)
Homo sapiens	131 pounds	1,350 cubic centimeters (present–300,000 years ago)

Australopithecus was an early biped who lived about four million years ago. His brain was fractionally bigger than that of a chimp. Like most chimps, he had a large torso and proportionately long arms. Not yet fully adapted to the arduous treks that our species has often

---

1   This is based on the average weight of *Homo sapiens* in recent prehistory.

undertaken, he stood not more than four feet tall. From his brain size, we can infer that he was smarter than a chimp—but not much.

Then, right at the beginning of a geologic age called the Quaternary, the earth suddenly grew colder. This was about 2.6 million years ago. Later, we'll discuss why this is thought to have altered human evolution. What's notable is that soon afterwards our ancestors' brains tremendously expanded, and the first creature we call *Homo* appeared: *Homo habilis*. The name comes from a Latin expression meaning "handy man," and that reflects his ability at making tools and his invention of fire. Human brain size grew by almost 50 percent. In fact, adjusted for weight, it probably expanded by more than that as *Homo habilis* took on a form that was designed for incessant walking. This meant he had shorter arms and a less muscled upper body. This was the first great wave of human evolution.

The second great wave happened over the last two to three hundred thousand years. Before that, two million years passed, and a large number of new hominids appeared. Their brain size varied. But if you adjust for their increasing size and growing stature, there doesn't seem to have been that much of a shift. Things did change dramatically, however, around the time that *Homo sapiens* appeared. This transformation wasn't limited to *Homo sapiens* as other hominids like the Neanderthals and the Denisovans also emerged. They were alike to us in that they had brains that were 30 to 50 percent bigger than those of earlier hominids. In this way, they were different from the various types of *Homo erectus* whose skulls have been found in parts of Asia and Africa.

The first *Homo sapiens* evolved in Africa. Sometime later we crossed the land bridge from Africa into Asia. With this, we began our conquest of the globe. That placed *Homo sapiens* in competition with some of those larger-brained hominids present in other regions, like the Neanderthals in Europe and the Denisovans in East Asia. Eventually we interbred with these hominid subspecies, and small amounts of their DNA survive in present human populations. For the most part, though, it seems that we displaced them.

As this was happening, lasting human art was created, and the earliest burials took place. That was in the last 100,000 years. The change becomes particularly noticeable in excavations of sites dating from roughly 75,000 years ago. Was this flowering of recognizably human activity connected to the discovery of spoken language? And was this something unique to *Homo sapiens* that explains our abrupt rise? Geneticists who have looked at one of the genes for spoken language, the FOXP2 gene, have shown that it regularly turns up in the *Homo sapiens* genome by 125,000 years ago.

That's not a minor point. As we have mentioned, when the version of the FOXP2 gene found in humans mutates or is absent, people cannot speak. That's demonstrated by a family in England. All of its members have a defective copy of the gene, and, though they have respectable IQs as measured by nonverbal tests of intelligence, none can talk. Without the FOXP2 gene, there is no spoken language.

Genetic tests have provided some—albeit disputed—evidence of a variant form of FOXP2 in the Neanderthal genome 260,000 years ago, and paleontologists have argued that the shape of the Neanderthal hyoid bone indicates that he was probably capable of speech. These last two points have been taken as indications that *Homo sapiens* was not alone in possessing spoken language and that this cannot explain our triumph over the Neanderthals and the other extinct hominids. But versions of the FOXP2 gene are common in most mammals and many birds. All these variants are involved in oral expression. Yet, so far as we know, only the one found in *Homo sapiens* works to make spoken language possible—not grunts or warbles. What the Neanderthal version of FOXP2 did we do not know. Moreover, the presence of a variant form of the FOXP2 gene in Neanderthals doesn't even prove that this didn't come originally from *Homo sapiens* as the two groups mated, and much Neanderthal mitochondrial DNA, which comes exclusively from the mother, seems to be inherited from a *Homo sapiens* female who lived before 200,000 years ago.

On the other hand, if we assume that the appearance of the FOXP2 gene set the stage for human speech, it explains many other things.

It isn't apparent otherwise why *Homo sapiens* should have displaced the Neanderthals and Denisovans. Without speech, *Homo sapiens* would seem to have been at a disadvantage, especially with regard to Neanderthals. Neanderthals were more muscular, and they had equally big, if not larger, brains. But they were barely hanging on in their traditional territories. Study of the bones of one clan reveals a young Neanderthal whose parents were half-siblings. Close to starving, he was from a group that had become greatly inbred—even before the incestuous union of his mother and father. These are the mating patterns of an animal approaching extinction. Then, quite suddenly, his speaking rival appears, and soon afterwards *Homo sapiens* rapidly multiplies. More, we flourished, producing astonishing art and inventing new modes of life. The second great wave of hominid evolution had taken hold. Is that a coincidence?

A nearly definitive proof that Neanderthals lacked spoken language comes from an examination of his tools. Archeologists have found that whether they are looking at ancient Neanderthal sites in France or Russia, the implements are nearly identical. Nor does their fabrication improve over time. The sharing of knowledge and the existence of a distinctive local culture of production that we associate with *Homo sapiens* are missing. Only one explanation would seem to explain this: Neanderthals couldn't pass along new ideas by talking. By contrast, *Homo sapiens* were rapidly working out and passing along designs for novel weapons, like bows and arrows and spears made from bones. That Neanderthals weren't able to follow the example of humans who were employing bows and arrows is particularly telling.

Neanderthals must have seen *Homo sapiens* using the bow and arrow as, living right alongside them, we employed it for thousands of years. Yet they never crafted them. More than likely that was because their manufacture is a complex process that virtually necessitates the use of language. Think about just a few of the steps involved in producing a prehistoric bow and arrow. We had to master the art of splitting off choice strips of sinew, which we used to make the bowstring. We had to learn which woods had the right mix of strength and elasticity and how to dry and hone these. We had to train ourselves in the process by which a notch is set within the shaft of the arrow so it can be fitted to the bowstring. We had to practice and take instruction in crafting the arrowheads and shooting the arrows. Finally, we had to figure out how to find the ingredients for resins and then become expert in their formulation as these were required to produce the glues needed to bind the separate parts of our new weapons together. Then, once we acquired this knowledge, we had to have the cooperative instincts to use it effectively, working in groups, out on the hunt. This is but one of many examples of the way in which the invention of speech set us out on a contrasting path of discovery, one that would take us on our route from prehistory to the internet.

The development of spoken language compelled the evolution of a more sociable species, and these changes worked in tandem, mutually reinforcing one another. That's suggested by another clue about what was happening. Around the time that we started making art and burying our dead, we appear to have become more docile. So, while it's conceivable that Neanderthals were perfecting language when *Homo sapiens* arrived, our increasing tameness would have made us more capable than they were, whether we were busy fabricating new weapons, hunting or fighting. For when speech is combined with high intelligence and a dutiful nature, you have a fearsome animal—much more than a rival constantly warring with his own kind. And the heavy brow ridges and the stocky build for which the Neanderthals are known indicate that they had much higher testosterone levels than humans. That does not suggest an especially manageable creature.

Evidence for this transformation comes from within our mouths. In the last 30,000 years, our teeth and jaws have been getting smaller. That indicates that we have been becoming increasingly tame. Still, it's likely that early *Homo sapiens* were not that much different than we are. We are the same species, after all. Another clue that we have not changed so much is that our level of sexual dimorphism stabilized at its present level 300,000 years ago. This tells us that monogamous human mating patterns were established long ago.

If the *Homo sapiens* who lived 100,000 years ago were alike to ones who are alive today, this would explain our willingness to follow leaders who led us into new terrains, regions that were far away from the places where our tribes had hunted and foraged. In many respects this decision to leave Africa was curious. As there were so few people at the time, it seems unlikely that this was something that population pressures dictated, and in departing, we were moving away from the herds of game animals and the spots for digging up roots and finding seeds that we knew. Over and over, we headed into the regions where other hominids lived. These areas were not only populated by potentially hostile competitors. Often they had brutal weather and harsh landscapes. We advanced into the Arabian deserts and the Caucasian mountains, the Siberian tundra and East Asian tropical forests. This willingness to undertake these expeditions was likely not a sign of courage so much as it was an indication that we were tractable—easily piloted.

So now we know when we became domesticated and when our brains expanded. What this does not tell us is *why* this happened. To understand that we need first to look at the evolutionary history of other cooperative hunters: creatures that went through parallel transformations. In that way, we can begin to grasp the reasons our brain has undergone a process of convergent evolution with dolphins and whales.

Most marine mammals are members of one of two orders: cetaceans or pinnipeds. There are other aquatic mammals—polar bears and otters, for example—but the greater number come from among this pair. As such, most people assume that they are closely related. In fact, their common ancestry is quite remote.

The first order includes porpoises, dolphins, and whales, while the pinniped order is composed of seals, along with sea lions and walruses. There is an additional subbranch of the pinniped order made up of creatures known as sirenians: sea cows and manatees. All these creatures have large brains and are self-controlled and intelligent.

The ancestry of the cetacean order connects them to even-toed ungulates (e.g., deer, horses, camels). The closest living relatives to the pinnipeds are bears. But, no matter what their descent, all aquatic mammals had to evolve in similar ways in order to survive in the water. This presents seemingly insuperable difficulties for mammals as we lack for gills. In response, both the cetaceans and the pinnipeds developed a range of highly specialized mechanisms to overcome the fact that neither can breathe in the seas and oceans in which they live. However, unlike cetaceans, pinnipeds don't echolocate; they lack the ability to throw off noises and then determine what's around them, using a form of sonar, as cetaceans do.

What both categories of aquatic mammals did develop was big, highly folded brains. Neurologists refer to the way in which ridges and valleys form in the brain as gyrification and sulcification. These wrinkles provide extra space through which connections within the brain can form. So highly folded brains increase speed and efficiency for performing complex mental tasks. You can see the result in the included pictures. Because of this, cetacean and seal brains are all unusually folded—as are elephant brains. Here are a few examples.

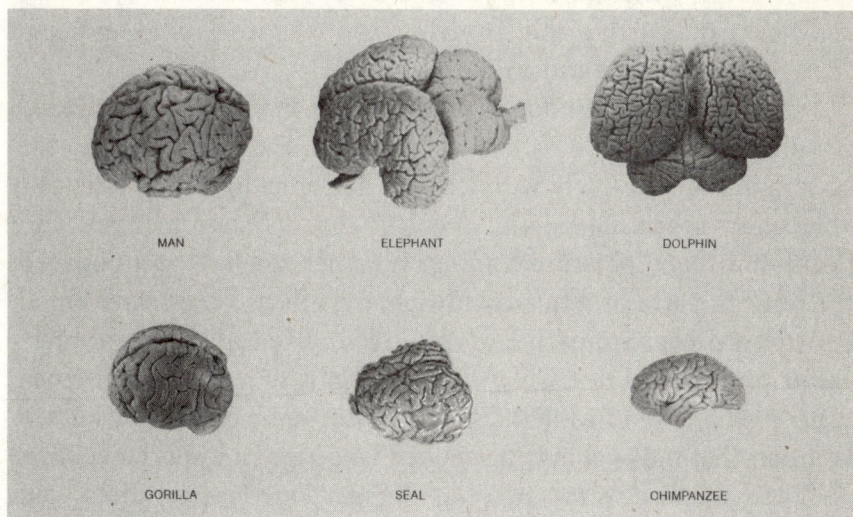

MAN   ELEPHANT   DOLPHIN

GORILLA   SEAL   CHIMPANZEE

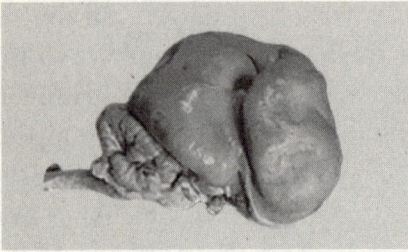

Manatee brain (above).

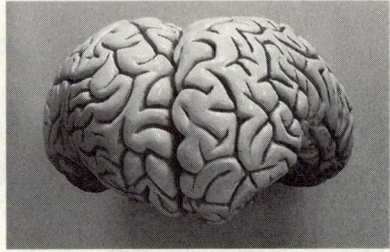

Blue whale brain.

You will notice one exception among the marine mammals. That's the sirenian brain. While it's large, it appears to be a big, disorganized heap of gray matter. It may be worth pointing out here that manatees and sea cows have a completely different diet than other pinnipeds and most cetaceans. Seals and dolphins are carnivores. Seals consume everything from sardines to lobsters. Dolphins are hunters as well. This is also a crucial difference between humans and nearly every species of primate. As we have mentioned, while chimps occasionally eat meat and they value it highly, they don't take much of it in. How and why we switched to eating flesh is a key part of the story of how we developed, evolving and surviving, and why we wound up being different from chimps and gorillas.

When you go to your doctor for a checkup and he takes your blood pressure, he may ask you a curious question: Have you eaten in the last hour? You will likely wonder why he wants to know. The spur to the query lies in the amount of energy that digestion requires. This affects your blood pressure reading. Whether you have just polished off a bowl of pasta or a tuna sandwich, the effect is the same: Blood flows toward your stomach and intestines, lowering the blood pressure in your extremities. Great as this effect is in humans, it is much greater in chimps and gorillas. That's a consequence of the only human invention that ranks alongside spoken language in importance: fire.

When we discovered fire, we changed our bodies, diets, and modes of life. This transformation took place during that first great wave of human evolution, and it's probably not an accident that many of the most dramatic shifts in the human fossil record occurred at the same time. Because we can cook our food, we expend less energy in digesting it. And since we need to expend less effort in breaking down our food, humans have smaller stomachs and large intestines than chimps and gorillas. This means we have more oxygenated blood and blood sugar left with which to feed our brains, and they require a lot of it. In our resting state, our brains use 20 percent of our blood glucose. Scientists refer to this as the expensive tissue hypothesis, and, as a further proof of it, it's been found that our blood proteins have evolved to give still more of our blood glucose to our minds. By contrast, chimp and gorilla blood proteins send more of their glucose to their muscles. It's believed, in fact, that this may partly explain an ape's much greater strength.

These are not the only important changes that result from the invention of fire. Eating became more of a communal activity as we came to consume our meals collectively by the hearth. And fire changed our diet since cooked food can be eaten when it's less than perfectly fresh. That means animals can be killed and then consumed hours or even days later. Inadvisable and disgusting though this may be, it means we can eat scraps of animals killed by other creatures. Altogether, fire permits us to digest much more meat than chimps and gorillas. So we rely less upon the consumption of leaves and stems. The core of a gorilla's diet became an occasional component in our salads. Moreover, the process of cooking set us out on the route to learning about curing and preserving foods since we employ fire in the process of smoking meats and cheeses. With all these changes, humans came to depend more heavily upon hunting—and eventually livestock raising—as a source for food. Because these activities require collaboration that further pushed us in the direction of sociability.

Yet the changes that took place in our diet may not be immediately apparent from our dentition. Like chimps and gorillas, we have twenty infant and thirty-two adult teeth. The most obvious dissimilarity that you can observe between our teeth and theirs is that chimps and gorillas have much bigger teeth, and the canines in their upper jaw are more pronounced. Since they eat little or no meat, this is counterintuitive. However, if you think about it from the point of view of behavior, it starts to make sense. As chimps and gorillas are wild animals, they have prominent canines, which are not meant for eating food but are employed in order to send a message to other chimps and gorillas. They have display value and are of use in fighting. Chimp canines can be as much as four centimeters long—more than an inch and a half. Gorilla fangs can be up to two inches long, even though they eat no meat at all.

Here again we see how different people are from chimps and gorillas. Where humans consider it to be utterly debased to bite another person, chimps and gorillas depend upon these teeth for intimidation and for doing injury. When chimps murder other chimps, they often chomp off their rival's fingers and toes. They also do this to primatologists. Indeed, one government study found that more than two-fifths of all primatologists report that they have been bitten.

In any event, let's return to humans. Confirmation of the theory that humans shifted to a diet based on meat right after the Quaternary Period began has been provided by paleontology. It turns out that we started butchering flesh by at least 2.5 million years ago—within 100,000 years of the change in the climate caused by the beginning of the Ice Ages. But research on prehistoric fossils has also revealed something unseemly. It appears that early species of humans weren't just hunting live animals We *were* eating meat that was days old and infected with masses of deadly microbes. These migrated into our bones, damaging our innards. Essentially, it seems that we were acting as vultures, eating carrion. That was to be expected. Why? One thing that primatologists rarely mention when they discuss the supposedly great kinship between men and apes is the geographical distribution of primate species. The total number of primate species is a subject of dispute as what a species is and how narrow our classifications should be are matters of argument. At the low end, there are experts who say there are just 89 primate species. Most suggest that there are more than 500. Whichever the case, there are no more than half a dozen living anywhere other than the warmest regions of the globe.

Chimp teeth.

Gorilla teeth.

This contrasts with reptiles. Because snakes and lizards are cold-blooded animals that take much of their heat from the sun, they are also better adapted to the tropics. Nonetheless, a half dozen lizards are native to the Cascade Mountain chain of Washington State and Southern Canada. But almost all primates are found in trees within tropical forests. This is where they spend most of their time, waking and sleeping. Not one primate is native to the United States, Canada, Europe, or Australia. All chimps, bonobos, and gorillas live in equatorial Africa. Humans, of course, travel ceaselessly and socialize continually, and we can be found in all parts of the globe. So what happened?

The general belief is that the change to a carnivorous diet took place when humans left Africa's tropical forests at the start of that Quaternary period. These were the first of the succession of Ice Ages that have occurred throughout the epoch of hominid evolution. This would explain why we had to adapt: As the forests in which we had been living disappeared, we moved out onto grasslands in which game animals grazed, regions with many fewer trees and leaves. This should remind us of the reason why nearly all primates live in tropical forests. It's where the greatest amount of vegetation is. Thus, it's where a vegetarian with skills at climbing would want to live since the opposable big toe that all primates possess assists in tree climbing. That makes it easier for primates to reach those leaves and to escape would-be predators.

Wolf's mona monkey: Typical primates, they live in trees of tropical forests.

The mandrill, a common monkey species—note that the animal is not adapted for hunting. His long arms hold branches.

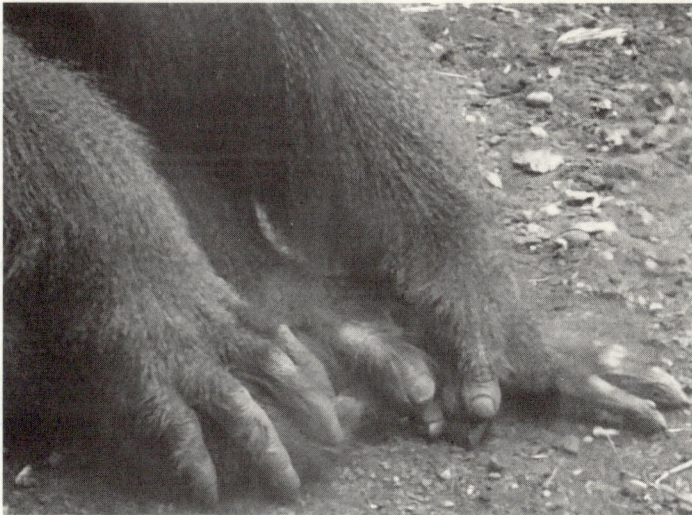

The mandrill's prehensile feet. Photo credit: Jonathan Leaf.

Note the gorillas' long arms and prehensile feet, which curve around and grasp the tree branches. While they prefer to knuckle-walk on the ground, in the dry season they climb trees to look for leaves.

The one genus of primates that is not dependent upon a fruit and vegetable diet is the tarsiers. At one time, tarsiers lived throughout Asia. Today, the only extant species live in tropical forests in four countries in Southeast Asia. Their bodies are rarely more than six inches long, and their diet is composed not of meat but small insects. Nocturnal creatures, they have eyes that are bigger than their brains. This is the sole evolutionary road open to tree-climbing primates that wish to survive by relying upon the consumption of something other than fruit, leaves, and stems. That's because there are no big-game creatures wandering about in the trees of tropical forest. Likewise, this is why the chimp's limited hunting is focused upon colobus monkeys—and other chimps. The colobus monkeys are the only substantial meat to be found in the trees of Central Africa that won't fly away, other than the chimps themselves.

This should tell us that our shift to living outside tropical forests must have been nearly impossible at first. We certainly were primates when we left the forest cover, and primates are not well-suited to living outside of tropical forests or to hunting for game animals. Part of the problem is that we have stereo-optic vision. In plain language, this means that our eyes face forward. That's an advantage in terms of depth perception, but because of it we lack for peripheral vision and cannot see potential predators. (It's no accident that nearly all animals that graze in open fields have eyes on opposite sides of their heads. These yield them a wide field of vision and the ability to spot attackers from both sides.)[2] Nor do we have the speed and agility of big cats or dogs or their powerful jaws and sharpened claws. A spotted hyena's bite is seven times as forceful as ours, and, as we all know, bears and jaguars have claws so steady and strong that they can use them to climb trees. All things considered, it would seem that we would be complete failures as hunters. Rather than the hunters, we ought to be easy prey.

That's consistent with what we know of chimps. When they wander about on the ground, they became targets for leopards and lions and other animals with the natural armature for predation. As violent as chimps are, they are no match for these big cats, and even in groups they are in danger for their lives. Primatologist Christophe Boesch followed chimps in the Tai National Park in the Ivory Coast. His aim was to see how often they were attacked and what sort of dangers they faced. He reports:

> [D]uring the study period, individual chimpanzees suffered an estimated predation risk of 0.30 per individual per year with or without injuries and a mortality rate of 0.055 per individual per year due to predators. In other words, on average an individual will be

---

2   Because humans live in open fields and are subject to predation in a way that apes and other primates are not, we have a much wider field of vision than apes. Hence, our eyesight is much less stereo-optic. On average humans have at least twenty-one degrees more range of vision than any ape species. See Eric Denion, Martin Hitier, Eric Leviell, and Frédéric Mouriaux, "Human Rather Than Ape-like Orbital Morphology Allows Much Greater Lateral Visual Field Expansion with Eye Abduction," *Nature: Scientific Reports* 5, no. 12437 (July 2015).

attacked by a leopard once in 3 years and 4 months and will be killed within 18 years.

When leopards manage to catch a chimp, the results are ugly. Referring to the observation of one of his colleagues, Boesch comments:

> Rushing to the spot with the female and the alpha male, he saw Brutus, an adult male, encircle the body of a motionless subadult, some of whose viscera were visible. I arrived there at 8.17 hrs and could confirm that Tina, a 9-year old female, had been killed by a throat bite of a leopard (later examination of the skeleton showed that the leopard bite had broken her second vertebra). She presented 23 claw cuts all over her body, one cutting the skin of her belly and partly exposing the viscera, two others on the left side perforating her lungs.

On another occasion, Boesch states:

> I saw Sartre, a 10 years [sic] old male, giving these despaired calls while he climbed up into a big tree. He was staring to the ground right under him and hitting his hands on a branch in a half temper-tantrum, all hair erected. On the ground, his mother, Salome, laid motionless on the back. Eight claw stripes covered her trunk, whereas her left chest looked awful as the leopard had bitten her there, compressing all the ribs which now formed a blood-stained protrusion.

Other chimp researchers have reported lower levels of chimps suffering from attacks by big cats. This is less of a threat to the chimps in East Africa. Regardless, the perils that chimps are confronted with are magnified on open fields. That reflects the greater possibility of attack *and* of dying in a stampede of game animals. So, while it's hard to make a direct comparison, it's certainly apparent that prior to the invention of guns we were at unceasing risk for our lives once we departed from the tropical forests and lost the ability to hide by climbing trees and sheltering ourselves in their branches. That's made still more true because though we are more cooperative than chimps and better armed, we're also weaker. And, as it happens, it seems that on several occasions we almost went extinct.

That's indicated by tests of genetic diversity. Hard as it may be to believe, all eight billion humans possess less genetic diversity than the few hundred thousand surviving chimps. Those tests suggest that there have been "bottlenecks" in the past. That's a fancy way of saying that at least twice during the long period of prehistory the total number of humans in the world dwindled to a few thousand. Some geneticists even think that there may have been fewer than a hundred *Homo sapiens* left at one point.

Like pinnipeds and cetaceans, we were working against the obvious patterns of evolution. Just as they were trying to survive as hunters in the oceans though they lack gills, we were striving to keep ourselves alive as hunter-gatherers without big canine jaws and claws and little in the way of quickness, balance, or agility. So why and how did we survive and come to populate the globe?

Hunting takes more savvy than sitting in a tree grabbing at the leaves and stuffing them in your mouth. It requires stealth, patience, and planning. For pinnipeds, cetaceans, and humans, it requires collaboration, which is to say sociability. You can feel this when you spend time with a dog. That he is so present emotionally is a consequence of his ancestry as a hunter. Since he needs to work with other dogs to feed himself, he must be aware of what the other dogs in the pack are thinking and feeling in a way that a gorilla need not. It's no accident that hounds are especially social and emotionally present as their breeding has been for the hunt. Ironic though it may be, hunting animals who operate in groups tend to be docile. And the less they have in the way of killing tools—sharp claws and big fangs—the more this becomes necessary.

Among the most popular and influential of the volumes written about human ancestry was Desmond Morris's 1967 book *The Naked Ape*. Composed by a writer with a dual background in surrealist painting and zoology, it offered a number of important and original ideas. Among the most essential is its explanation of why primates tend to be so violent. Morris asserted that this was because almost all primates are vegetarian. He thought this was important as herbivores are less equipped with bodily tools for killing: the claws and the huge jaws that carnivores possess. His idea was that since they don't have those instruments for mayhem and can't so easily slay one another this meant that the brutal aspects of intraspecies competition were less of an ever-present threat to their survival. From this, he concluded that they faced less evolutionary pressure to curb their impulses toward grappling with one another. In this way, he argued, they had become more innately aggressive. Regrettably, what many people remember from the book is the word "ape" in its title, used as an identifier for humans, along with its array of offbeat opinions. Among these was his suggestion that human ear lobes had become enlarged as an erogenous zone meant to facilitate pair-bonding.

What many readers missed or forgot was Morris's observation about human instincts toward aggression: When we turned to hunting, we acquired deadly instruments, which forced us to be more cooperative with one another. That pushed us on an evolutionary path to being less violent. Morris was right about this. Primate researchers who have worked with chimps report that while they use rocks to break open nuts, they are unable to learn how to make flints, to engage in what is known as stone-knapping. The mental apparatus required to craft tools for hunting game and for cutting open flesh appears to be absent. But, as humans are hunters, we have spent millions of years finding and fabricating flints, and between 20,000 and 40,000 years ago we invented the atlatl. This is a tool for launching a spear at greater velocity. In order to make use of these discoveries we had to become tamer and better at watching, teaching, and learning the arts of tool-making.

This brings us back to the manatees and sea cows. The sirenian diet is based on consumption of huge quantities of aquatic vegetation: sea weed, kelp, sea grass, cord grass, and turtle grass. This does not require the skills or the cooperation entailed by hunting. To describe it as a relatively mindless activity is in some respect figurative. It is also literal. It shouldn't surprise us then to find that a manatee's brain, big as it is, is not folded. The wrinkles are absent. Unlike the rest of the pinniped order, the sirenians do not require a highly specialized brain or a particularly social nature. In this regard, we see the contrast not only between types of pinnipeds but also between humans and primates. The hunting species are more social, intelligent, and capable of learning.

We can see this within mammalian families. There is a pronounced difference in brain size and power between the toothed whales, species that actively seek out their prey, and those that merely filter their meals through giant mouths that act likes brushes. This second category, the so-called baleen whales, possess smaller, less active mental structures. You may have noticed how much more folded the beluga whale brain is when compared with the blue whale brain. That's because the beluga is a hunter, and necessarily belugas are highly communicative. In fact, they can imitate human speech, and they like to play with people by trying to do so. Hence, their brains are full of ridges: designed for constant thinking and interaction. By contrast, the blue whale is a creature that simply sticks its mouth out regularly in order to swallow masses of seawater, which it filters for food. Not a hunter, it is not as intelligent as the beluga whale, and it has a comparatively disorganized brain, as you can see from pictures.

This pattern also pops up in a comparison of the pinnipeds. That's why you can go to a show at the zoo and watch a seal or a sea lion perform tricks but not a sea cow or a manatee. The sirenians didn't evolve to be good at communication and coordinated activity. They didn't have to. You might also notice how much less folded the gorilla brain is when compared with the chimp's and then recall that gorillas are pure herbivores.

This pattern also distinguishes humans from rats and mice. Relative to body size, rats and mice have big brains, and they excel at solving puzzles. For example, they can rapidly figure out how to exit from mazes. However, they don't cooperate much as hunters, and their use of any sort of language tends to be modest. So their brains are likewise lacking in folds. Like sirenians, they use their intelligence to adapt to unexpected problems but not to hunt with other rodents. By contrast, we would not have survived without that ability, and the difference is apparent in even a cursory study of the shape of our brains.

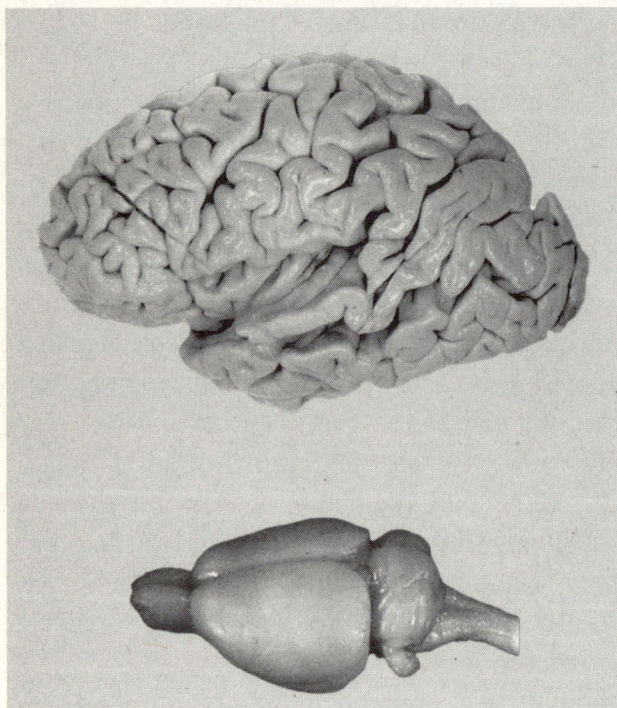

Human Brain

Rat Brain

We can also pick up clues about who and what we are by examining elephant brains. Though elephants are herbivores, not omnivores or carnivores, they are highly social and tame, and they communicate with one another orally and through a wide range of gestures. Most of their calls are made at very low frequency—roars so low that they are not audible to humans. They have dozens of these, and they can be heard up to six miles away. Each appears to have a specific meaning. These are used to protect the troop and to tell each individual what his role is. As elephants have an incredibly keen sense of smell, they gently touch each other's trunks to test each other for changes in scent. They can also pound the ground with their hooves, producing sounds that are detectable as seismic vibrations. That serves to provide another means of distant warning to members of the troop who are at a remove. And subtle and capable communicators that they are, elephants, like people, have highly folded brains.

However, one animal stands out most for its wrinkled brain. Dolphins are the only creature whose cerebral cortex is more folded than that of humans, divided up and constructed for its separate learning tasks. Undoubtedly, this is because, like humans, they must engage in collaborative hunting to survive—and they have forms of spoken language. These have permitted them to overcome their equally peculiar competitive struggle. Just as we were able to adapt to the role of hunter-gatherers even though we have stereo-optic vision, little speed, flat nails, and small teeth, they survived as hunters in a world in which they cannot even breathe. In both cases, the key was the union of high cooperative intelligence with sociability and language.

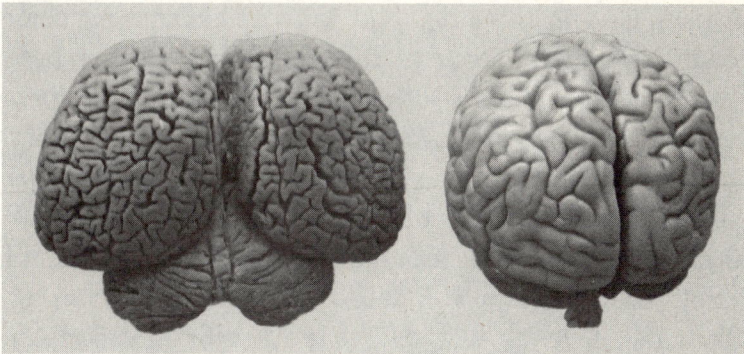

Dolphin brain versus human brain.

In our case, the development of higher intelligence was aided by a chance mutation. As people came to consume more meat, humans gained an improved version of what is known as the apolipoprotein (apoE) gene. Chimps are without this mutation, and its absence places them at increased risk of bacterial infection when they eat meat that is infected with microorganisms. The addition of the apoE gene allowed us to survive the effects of contaminated meals, and, when this was combined with cooking, we became more able to digest meat that was less than freshly killed.

This was not the only benefit of the mutation. Not only does it protect us against the immediate risk of food poisoning, but it prevents longer-term effects of microbial invasion. Without this, we would suffer from cardiovascular inflammation, which would gradually destroy our hearts and brains. Our brains would be more susceptible to early-onset Alzheimer's disease, and our bodies would suffer youthful heart disease. In this way, the apoE gene extends our lifespan. Lacking this, chimps that eat rotting meat wind up suffering from both senility and damage to the cardiovascular system.

And, as it happens, we have a compelling proof of how powerful the apoE gene is and how it led to different evolutionary histories for chimps and people. Primatologists have come to realize that chimps never eat carrion. If they have not killed an animal themselves, they will not consume its scraps. Referred to by animal researchers as kleptoparasitism, it's something that both lions and hyenas regularly do. Granted the chimps' desperate yearning for meat—one that prompts them to eat their own babies—it's striking that they and every other primate species will never engage in kleptoparasitism, but humans do.

What's more, the apoE gene has additional uses. As a side benefit the apoE gene increases the amount of cholesterol reaching our skulls, and, as the myelin sheathing around nerve cells is composed of fatty tissue, this boosts the supply of needed brain matter, increasing our overall mental efficiency. The apoE gene also offers a striking example of convergent evolution. Independently, the mutation appeared in cetaceans, and it has been found in bottlenose dolphins and killer whales at almost exactly the same blood concentrations as it exists in humans.

Through all these changes, we became more equipped for a meat-based diet and more capable as hunters. That increased the amount of nutrition we could take in, and it expanded our potential territory to the entire planet, even as chimps and gorillas remained confined to the rainforests of central Africa.

Passing through these changes, we also gained in our instincts toward courtesy. Manners became an increasingly important part of our nature. In order to avoid conflict with one another—now equipped with deadly weapons—we had to evolve toward complex rites of politeness. The dramatist George Bernard Shaw observed in his play *The Devil's Disciple* that it is a characteristic of people that we show greater courtesy toward our enemies than our friends. This serves as a striking contrast with the patterns of behavior of chimps and gorillas. They are most likely to snarl at their foes. All this explains some of our great similarities to carnivores. Why then in other ways are we so different from them?

CHAPTER 5

WHY WE ARE DIFFERENT FROM MOST CARNIVORES

CARNIVORES LOVE SLEEP. THAT'S ESPECIALLY true when they are apex predators. Male lions are typical. They are dead to the world for up to 20 hours each day. Primates also tend to be fond of slumber. At the upper limit is the three-striped night monkey. Found in the Amazonian rainforest, he drowses for an average of 17 hours each day. Two species of macaques, gray mouse lemurs, ring-tailed lemurs, and cotton-top tamarins catch more than 15 hours of shut-eye. At the low end of the spectrum are chimpanzees. They take in between 9.67 and 11.5 hours.

So, as humans are meat-eaters and we are supposed to be primates, you might assume that we would be superior sleepers. But here's the curious thing: No primate whose nocturnal resting patterns have been studied spends less time at it than we do. While the range among primate species varies greatly, it is always far in excess of that observed among humans. As adults we typically get less than seven hours per night of sleep. We also defy another rule of great apes in our resting patterns. Chimps, gorillas, and orangutans build nests. We do not.

What explains this? A clue might be found in looking at what sort of mammals tend to function on little sleep. Once again, the answer is large-brained herd animals, such as giraffes, horses, and elephants. If they sometimes look a bit ragged at the edges, there is reason. Since these creatures must stay alert to avoid becoming prey, all of them sleep less than five hours each night. Were we to plot ourselves on a graph, we would fall somewhere between the primates and the ungulates. This makes sense when we begin to think about how we live.

Lions and tigers have the bodily tools to easily kill any creature that might threaten them, and most primates can climb trees to escape predators. For most of our evolutionary history we have been more like giraffes and elephants. Our evolution has arisen with an ever-present fear as our physiology and anatomy are different. We have had to be awake and alert. Our collaborative instincts have been employed not only in hunting but in jointly protecting ourselves from attacks.

Wandering off alone meant likely death. That fact has compelled us to develop a certain delicacy in our manners and a high degree of social engagement. Although we have evolved to hunt and to consume meat, we are not obligate carnivores. This is a term for an animal that depends exclusively upon flesh as a source of its nutrients. These creatures are within a category referred to as hypercarnivores. It is a select designation. Among mammals it is reserved for a small number, including big cats, some wild dogs, seals, sea lions, polar bears, and dolphins. Zoologists distinguish these animals from mesocarnivores or facultative carnivores. Most animals within the carnivore order are of this latter type. They have a diet that is mostly made up of meat, but on occasion they will also eat insects, mushrooms, fruit, and grasses. This is how skunks, foxes, raccoons, and mongooses feed themselves. It's worth noting that the hypercarnivores that hunt in packs are among the tamest animals. When we place dolphins and sea lions in zoos, we are setting them in environments whose territory is just a tiny fraction of what they have been accustomed to out in the ocean. Yet they will follow our directions, learning tricks and performing for us in return for treats. Similarly, dogs have proven to be the most easily domesticated and most affectionate of all pets. Mesocarnivores, however, tend to be less docile.

Until now I have not said how much meat *Homo sapiens* have typically eaten during our lengthy process of evolution. It is not unusual for present-day hunter-gatherers to take 30 percent of their calories from meat. Long ago though it was much more. In some cases, it was 80 percent of the total. According to a study of 229 hunter-gatherer societies, just 14 percent of prehistoric tribes derived less than half of their calories from animal sources. On average our ancestors were taking in twenty times as much flesh as chimps and infinitely more than gorillas and all other primates. The 55 to 65 percent level that is thought to have once been average for humans is several times as much as black bears consume. It may be five to six times as much. This amount of meat consumption places humans alongside lynx, bobcats, and otters as mesocarnivores.

That we became hunters before we learned the arts of livestock raising is one of the most extraordinary facts in the history of evolution. Creationists like to question evolution by pointing to the eye. Because its function involves the coordination of so many different parts, sight is remarkable. But one might argue the fact that humans learned to hunt is more startling.

To get a further sense of this, let's look at the mammal order as a whole. One difference between warm- and cold-blooded creatures is that while most fish and reptiles are carnivores, the vast majority of mammals consume plants. This is a consequence of the high rate of mammalian energy consumption. Maintaining our high body temperature requires enormous amounts of fuel. Mammals constantly burn huge quantities of it, and, since this energy ultimately comes from plants, there must always be several times as many mammals eating plants as there are carnivorous mammals consuming those herbivores. Were this not the case, meat-eaters would quickly run out of prey and die out as the original source of energy for animals is the plants that use chlorophyll to turn the sun's energy into edible sources of power: sugars, fats, and proteins.

To put this in concrete terms, think of a big cat. Male lions eat between ten and twenty pounds of meat each day. That requires a lot of prey. During the course of a year, a single male lion consumes between fifteen and thirty-five animals the size of an adult antelope. Since there are certain inefficiencies in digestion and a loss of energy in the transfer of energy from plants to the animals that digest them, a great many gazelles and antelopes have to chew up a huge swath of grass for that one lion to feed himself.

Consequently, of the more than 5,400 mammals, just 279 fall into the order of carnivore, and many of those are actually mesocarnivores or omnivores. Further, within the order, there are just nine families that live on land. These are the canids (dogs and wolves), felids (cats), ursids (bears), procyonids (raccoons), mustelids (weasels, badgers, otters), mephitids (skunks and stink badgers), herpestids (mongooses), viverrids (civets and genets), and hyaenids (hyenas). What they have in common are big teeth with glaring canines, powerful jaws and chest muscles, sharp claws, four legs, and a highly developed sense of smell. Each has a purpose and a function, and it may be worth reviewing them since, as we have mentioned, we don't have any of those.

The teeth and the canines, along with the mighty jaw, are for tearing flesh apart and for striking at key veins and arteries, like the carotid and the jugular. The outsized chest muscles assist with pouncing on and pinning down the prey. The claws are not like our nails. They are not cuticles. They pass through the paws, connecting to bone and attaching to tendons. They aid in ripping up and grasping a victim, and they can be employed in climbing. The quadruped form is ideal for speed, balance, and agility, and the carnivore's enhanced olfactory organs provide it with an innumerable number of clues that can be used in tracking and hunting.

ANATOMY OF CAT'S CLAW

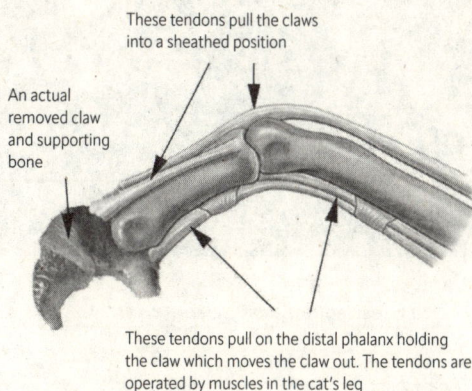

These tendons pull the claws into a sheathed position

An actual removed claw and supporting bone

These tendons pull on the distal phalanx holding the claw which moves the claw out. The tendons are operated by muscles in the cat's leg

Anatomy of cat's claw. Created by Michael Broad.

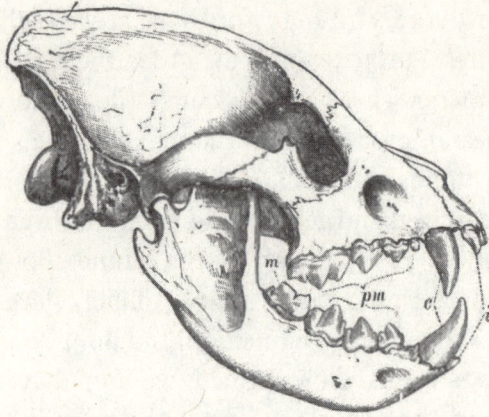

Hyena skull and teeth.

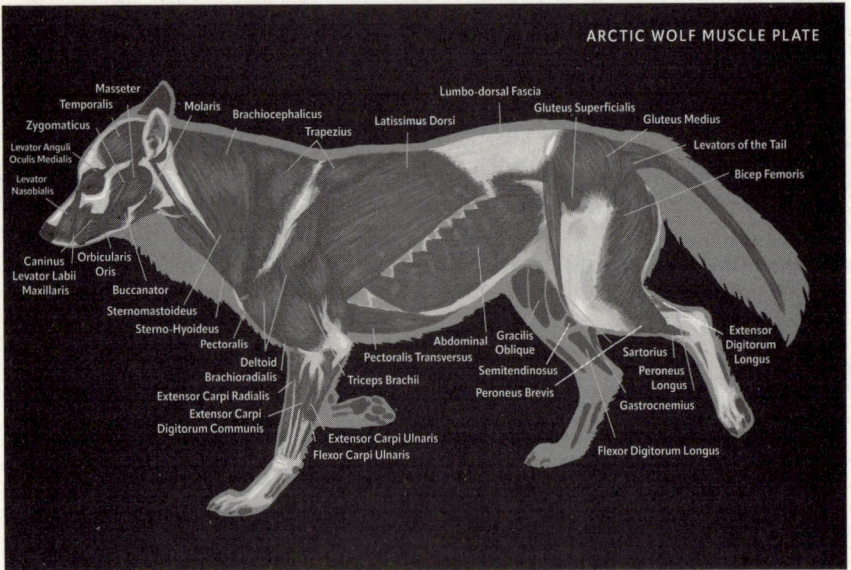

Wolf musculature. Credit: Sarahjane Bernhisel.

Weasel nose.

Cheetah legs.

Without these, we're like Muggsy Bogues. For those not familiar, Bogues is a former professional basketball player. He played in
the NBA for ten seasons, and he ranks in the top twenty-five players
all-time in assists. Yet he stands five-feet-three-inches tall. He made
up for that with a number of astonishing skills. He had amazing
quickness and strength. Those matched to extraordinary peripheral
vision, anticipation, and hand-to-eye coordination. And he played
with uncommon passion and energy. For all that, he's still a very rare
exception to the rule that short men can't excel at basketball.

That humans learned to hunt is equally remarkable as we have
none of the obvious requirements for the task. Evolutionary biologists often point to our use of tools as a substitute, but advanced tools
appeared tens of thousands of years after the first humans were born.
A hundred thousand years ago *Homo sapiens* could do no more than
attach sharpened pieces of bone or slivers of rocks of two or three
inches in length onto wooden poles. Consider that even with a hand-
gun it's exceedingly dangerous, acting alone, to try and approach a
stampeding herd. Yet we know from archeology that by the end of the
last ice age humans were hunting mastodons and woolly mammoths.
Nineteenth-century big-game hunters found that it could take up to
thirty-five shots with a regular gun to bring down an elephant. For
this reason, they came to use giant-bore weapons known as elephant
guns. Most notorious was the .577 Black Powder Express. These rifles
carried a three-inch long cartridge weighing more than four ounces.
In the present day, poachers will first poison elephants with cyanide
and then fire at them with automatic rifles. So how was it possible
for us to avoid getting killed hunting woolly mammoths when our
weapons were spears, bow and arrow, and, at best, pikes? Present-day
African elephants are enormous creatures. The males can weigh 13,000
pounds. But wooly mammoths were even bigger. Some weighed up
to 20,000 pounds, and their tusks were up to fifteen-feet long. To slay
them we had to evolve so that with great patience we could engage in
extremely close collaboration. Remarkably, we then learned how to
kill the mastodons and wooly mammoths so efficiently that we drove

them to extinction. The meaning is clear: To survive as hunters, we needed to undergo a profound change in our nature, and development of speech was almost certainly part of that.

Bear in mind, too, that even with the natural tools of the carnivore—the big teeth, the sharp claws, the speed, the keen sense of smell, and the powerful chest muscles for pouncing on a targeted creature—few predators attack animals that are as large as they are. If you have a housecat, you may know this already as while housecats are masterful at killing mice, they won't pursue rats. The predators that slay animals bigger than they are fall into two categories. Either they are exceedingly patient hunters like ermine and Komodo dragons, or they are stealthy pack animals that work together closely and hunt as a group. That's an apt description for hyenas, wolves, killer whales, and dholes (a species of African dog). Because humans were pursuing animals that were one hundred times our size and doing so without the natural carnivore equipment, we had to combine those traits and then make use of speech in addition. We had to evolve so we were capable of being even more cooperative than wolves and hyenas, and more patient than Komodo dragons. Simply put, we had to stop thinking and acting as primates generally do.

That defied probability. That isn't only because we were attacking stampeding behemoths that could easily crush us under their feet or fatally maim us with their tusks. There's even more that's rare and distinctive in this. It's exceedingly unusual for an animal to survive after it radically alters its diet. Scientists have long known that diet and behavior are related, and, as we have just observed, eating patterns, digestion, and skeletal structure are linked too. That means that an animal can't just change *one* of these. He has to change all of them more or less at once while arriving at an evolutionary design that satisfies the requirements for his survival. This is nearly impossible. However, that's what we succeeded in doing.

How difficult is it? Here's a list of the well-known animals that fundamentally changed their diet, going from herbivore to carnivore or the reverse, which are not extinct:

1. Bees. Bees descend from carnivores, like their wasp cousins, though they now feed off the nectar of flowers, which they pollinate.
2. Pandas. The ancestors of pandas were like all other bears. However, they now depend upon bamboo shoots for most of their diet, and they consume almost no meat.
3. Cetaceans. As we have noted, whales and dolphins are related to even-toed ungulates and are descended from herbivores.
4. Humans.

That's it. And all these animals went through far-reaching changes in physical structure, social structure, and behavior. As we all know, bees are relatively docile creatures that make honey, while wasps are aggressive insects that do not. Bees live in hives, while wasps build nests and hibernate in the wintertime. Bees can sting just once, wasps many times. Bees are hairy, wasps smooth. Pandas are famously slow and ponderous, unlike other bears, and they don't hibernate. In addition, they have an extra "digit" in their hand, the radial sesamoid bone, sometimes called their sixth finger. They use this to grasp the bamboo shoots that they eat all day long. They have to constantly swallow mouthfuls of these as they offer them far less protein and calories than a meal made up of flesh. One could go through much detail in describing the multitude of differences between hippos and cetaceans.

What's important is that no one pretends that we can come to definite conclusions about the activities of bees or dolphins through the actions of their relatives. But that is what most primatologists insist: They can explain human behavior through the habits and instincts of chimps and gorillas. In fact, the claims of primatologists are even more questionable as we not only changed our eating habits but also our principal mode of communication.

That brings us back to the second half of the story of our divergence of the primate order: our invention of language. Here, as well, neuroscientists are revealing a fascinating and unexpected tale. Strangely, it may partly explain another great mystery of evolution: why some people are attracted to their own sex.

CHAPTER 6

HOW SPEECH CAUSED THE SECOND GREAT WAVE OF HUMAN EVOLUTION—AND WHY IT MIGHT EXPLAIN PREFERENTIAL HOMOSEXUALITY AND THE WAY WE RAISE OUR CHILDREN

LET'S STEP BACK FOR A moment. Assume that the standard explanation given by primatologists for what humans are is correct: We are apes. The evidence they have adduced for this dates back roughly a hundred years.

Most primate researchers are somewhat embarrassed by Robert Yerkes, the figure with whom their field originated. Yerkes was a racist and a eugenicist. In many respects he was also a poor clinician and researcher. A leading "expert" on intelligence testing, Yerkes helped the government develop questionnaires that were used for Army induction during the First World War. From the results of these tests, Yerkes "proved" that Southern and Eastern European immigrants were of innately low intelligence. One group that Yerkes's research especially seemed to indicate was feeble-minded was Polish and Lithuanian Jews. Yet within a generation this same group was gaining admission to colleges and universities at such high rates that schools began applying discriminatory standards to keep them out. This did not, however, prompt Yerkes to publicly renounce his earlier claims.

Yerkes did switch his area of research. In the 1920s, he visited Cuba and spent time at the home of a famous chimp collector, Madame Rosalia Abreu. Soon afterwards he wrote a book called *Almost Human*.

It was the first influential work based on observation of chimps to put forward the idea that chimps and humans were siblings, or, at most, kissing cousins. This conclusion and the reaction triggered by the book led to funding for further investigation of the subject and the establishment of a Yale University research center near Jacksonville, Florida. Now at Emory University in Atlanta, up until 2022 it was named in his honor: the Yerkes National Primate Research Center.

The center has been involved in a great deal of horrific research. Some of this was exposed by the 1974 Frederick Wiseman documentary *Primate*. Among the center's past activities were brain extractions. Because chimps are so dangerous, they are kept in cages. These are either to be found on the Emory campus or nearby in a more wild but fenced-off setting.

The center was one of the main places where studies were undertaken during the 1970s that aimed to demonstrate that chimps could learn human language. The center's advocates argue that the few hundred words—at most—successfully taught to a handful of especially bright and attentive chimps prove that they are as, Yerkes put it, almost human. But we are back to our problem of how good the evidence is. If this is the test of how much an animal is like a human, then we can see clearly what humans are: a subspecies of dog. After all, as we have already noted, dogs can master more words than chimps, and they can live in houses. Like humans, they are domesticated. They are even more like us in what they eat. Dogs, like people, will consume almost anything.

Chimps have much larger brains than dogs. The difference in brain size is even more pronounced with respect to cats. So why can we teach cats to use a litter box, but we are unable to toilet train most chimps? The primate anthropologist will likely respond by saying that humans are alike to chimps *but that we are domesticated*. This is somewhat like saying that bears are just like termites but that they aren't wood-eating insects. The distinction is so great as to render the prior statement almost meaningless. It is not a matter of how we are the same except for one thing.

We can see this in other ways. Take but one more such instance with dogs: They are alike to people in that they can be taught to follow another creature, or they will do so instinctively. This brings up one of the many ironies of primate research: Humans can be instructed to follow chimps and gorillas in order to observe them. This activity is undertaken in order to prove our kinship with them. But the researchers have seemingly missed the obvious fact that the chimps and gorillas cannot be taught to follow us. Why did Yerkes ignore the significance of this point?

In his earlier work as a psychologist, Yerkes spent much time investigating nonverbal intelligence. This is where the trouble starts. Humans are a speaking animal. Speech by its nature compels us to see the world in a different manner from creatures without it. The first great wave of human evolution was spurred by our need to cooperate in hunting. It made our survival possible but only barely. The second great wave occurred around the time of the invention of language. It further sped the pace of change.

When I was in college, Jacques Derrida's idea of deconstruction was one of the most talked-about intellectual concepts. It holds that language is imperfect and unstable. It says that expression through words can create "unmeaning." Like so many academic notions, this reverses matters. By this, I mean that what is notable about language is its amount of specificity and what this suggests about our species. Consider the word "thrifty," and think momentarily about all its synonyms: frugal, prudent, stingy, careful, chary, cheap, miserly, niggardly, chintzy, close-fisted, conserving, mean, parsimonious, penny-pinching, provident, scrimping, tight, unwasteful. Although these words have similar *denotation*, they carry distinct *connotations*. From these they put across varying implications—so much so that it has been said that there are no true synonyms. Each word conveys something different from each other. But how do we know this? Are we being attacked, or are we being praised when we hear one of these terms applied to us? We know because we have evolved to an extraordinarily delicate awareness for language. This enables us to pick up subtle cues telling us how these words are distinct from one another. Otherwise, we would not know which are pejorative, which are affirmative, and in what forms they are being employed. In the absence of spoken language, we have no need for this agility. Yet, as we are a creature gifted with speech, we cannot thrive unless we are profoundly attuned to those around us and the precise meaning of words people use and the relationship of these words to all the other ones. This requires an enormously refined awareness of the speaker, the words, and the context. It demands that our process of thought is social above all else.

This even goes beyond the words' connotative meanings as these may be understood in speech and writing and as the words can be subtly distinguished from one another. We gain further cues through word order and positioning and by expression. Consider an example from the play and movie *The Philadelphia Story*. At its most dramatic moment, the heroine's fiancé accuses her of wantonness. He responds to her claim that she has done nothing wrong in taking a midnight swim with another man on the night before their wedding by angrily declaring, "A woman should *behave herself*, naturally." This prompts her ex-husband, who wishes to remarry her, to defend her by saying, "A woman should behave herself *naturally*." The two sentences are identical, but through the shift in emphasis they carry entirely different meanings.

Matters are even further complicated by the many meanings that each word has and by the enormous number of homonyms (e.g., hear and here) we employ, the use of metaphor and simile and other figures of speech (e.g., hyperbole, apostrophe, metonymy), and even the rare instances of contronyms (words like "sanction" and "cleave" that can have exactly opposite meanings depending upon the context). Only a highly social creature can understand these distinctions.

Chimps can be tickled, and, while they lack our voice box and vocalize differently than we do, they will respond with something alike to laughter. This appears, in fact, to be a common trait among mammals. Hence, tickling induces analogous responses in rats, dogs, dolphins, and many other species. And while chimps can't do much in the way of tool manufacture beyond sharpening sticks, they can use found objects as tools. For instance, they can smash rocks they grab in order to break open shells. However, this may not be such a notable activity as simple forms of tool-making have been observed among many species, including sea otters and corvids (crows and ravens), elephants, finches, parrots, and octopuses. Thus, bottlenose dolphins use sponges to get fish free from crevices or from the sea-floor, and then they instruct their offspring to practice the same trick. Similarly, the degu, a rodent related to the chinchilla, can be taught the means by which it can twist around a rake to uncover sunflower seeds, and some crows actually travel about with their "toolkit." Even many insects, like ants, regularly make tools. Moreover, one of the most consistent findings about tool-making is that corvids are better at it than chimps are. Unlike chimps, crows can carefully and pre-cisely shape objects that they use to reach sources of food.

As we have already noted, primate researchers especially love to point to the cooperative pulling experiment and the collaboration dis-played by chimps in hunting as proofs that they are highly social and much like us. In actuality though the only human behavior observed among chimps and gorillas but not in other creatures is kissing. Yet chimps and gorillas don't kiss for the reason most humans do. They do not smooch one another as an expression of romantic love as they have no such concept. Kissing is simply a pacifying gesture, akin to a massage, which is applied to a potential foe.

Thus, the common claim that primates are "highly social" raises the question: *Compared to what?* You can get some sense of what primate researchers mean by highly social in reading their extended accounts of the elaborate process by which chimps engage in com-petition and form alliances within their troops. In these case studies, they tell us about the various chimps and gorillas, and they describe their quirks and their daily rituals. There are also detailed discussions of their acts of violent aggression toward one another and the efforts to charm that sometimes come afterwards.

As Goodall puts it, "although a male chimpanzee is quick to threaten or attack a subordinate, he is usually equally quick to calm his victim with a touch, a pat on the back, an embrace of reassurance." Goodall made this observation in describing the aftermath of an incident in which one of the male chimps she studied had savagely beaten a female in the troop. This was followed by his incessantly tickling her, striving in this way to win back her favor.

These accounts acquaint us both with the routine patterns of ape existence and with the manner in which their hierarchies are formed. What remains unclear is how this proves that they are any more socially engaged than pack or herd animals like dogs, hyenas, wolves, or, for that matter, dolphins or elephants.[3] Could a mix of fighting and playing not be observed among them? Goodall has claimed that the primary and essential reason why humans are more collaborative than chimps is simply that we have a greater intelligence. But is this true, and do her reports support this idea? Or is cooperativeness a trait that is found among humans, dogs, and ants but not to any particularly significant degree among apes?

Tamed animals are always less inclined to fighting with their peers than their wild ancestors. Domesticated horses fight less than wild horses, domesticated turkeys fight less than wild turkeys, and dogs fight less than wolves. Inevitably they work together better. That difference would apply to the same degree to modern humans and *Homo sapiens* who lived 100,000 years ago, and those changes are but a part of a much larger pattern of changes. However, even noting that leaves us short of an explanation for why humans are highly cooperative but apes are not. No primate is highly cooperative—unless you count humans—but many species outside the primate order are routinely so. Yet to say that one set of animals is more like another because they have similar dentition but to ignore this point is to place

---

3   Dolphins and whales that hunt cooperatively can be thought of both as pack and herd animals. However, it is probably more accurate to regard most of them as herd animals because the primary reason they herd together is for self-protection. However, killer whales (orcas) should more readily be thought of as pack animals as the first and foremost reason they herd together is to assist one another in their various predatory assaults.

great emphasis upon a relatively trivial matter over and above something of the utmost importance. But this is what the field of anthropology has been uncritically doing for half a century.

Psychologists refer to psychopathic people as *anti*-social personalities. This is what most chimps and bonobos are: lacking in empathy, highly manipulative, instinctively brutal, and persistently disobedient. If you were to point this out to a primate researcher, they would likely respond by saying that this is a term from outside their field and that it is therefore irrelevant. But psychologists adopted it and incorporated it into their everyday terminology because it is accurate and descriptive. By the standard of humans—and most other mammals—primates are not inclined toward the collaboration and constructive social action that we depend upon to survive.

What then is the argument that primate researchers are making in their insistence that primates are highly social and much like us? There are several ways to make a logical argument. One kind is known as abductive reasoning. This is the process of reasoning we employ when we say that if a creature walks like a duck and quacks like a duck, it must be a duck. Plainly, employing this type of logic, humans can't be said to be primates as we don't walk or sound like them. A more sophisticated line of argument is the kind used in scientific experimentation: inductive reasoning. We craft a hypothesis, then we test it. The experiment showing that chimps consider the opportunity to provide food to another chimp as inconsequential appears to rule out this claim that they are that much like us. So, as well, does the neuroscience showing the profound differences in our brain. (In a later chapter, we will have more to say about that with respect to the neurotransmitters that our minds rely upon and the hormones that guide our basic impulses.) Yet the most common sort of reasoning we make use of is deductive: We examine the facts, and we draw logical inferences from them. Here, too, though the assertion that we are primates falls apart as we do not have the patterns of behavior that they do.

This is not only because chimps cannot work together well or learn as many words as dogs. There is another evolutionary puzzle that crops up, which we must explain. Humans are one of the few species among which there are individuals who are preferentially homosexual. Many primates engage in homosexual acts. Chimps do so almost routinely. But this is a means of establishing hierarchy, preventing conflict, and defusing tension. By contrast, many humans are expressly interested in and attracted to their own sex rather than the opposite sex. Like suicide, this is something that appears to conflict with Darwin's premises as it does not assist an animal in passing on its genes.

Richard Wrangham has noted that the only mammal species in which exclusive homosexual attraction has been found is the sheep. In what other large mammal species are preferential homosexual ties commonly observed? Amazon and bottlenose dolphins, elephants, giraffes, and bison. It is intriguing to again notice that humans wind up in the company of large-brained herd animals. For the moment, however, let us focus upon sheep as more and more is now known about ovine homosexuality, and sheep are the other species in which it is so clearly a matter of preference.

About 8 percent of rams show no interest in mating with ewes. They are not the only gay sheep though. Some females are freemartins. These are masculinized females: lesbian sheep. This phenomenon of the freemartin isn't limited to sheep, and it was recognized long ago. We have reports of it dating to the Roman Empire. Freemartins have also been identified among cattle, pigs, and goats. Most freemartins come from mothers who bore twins. Usually, one is male. During gestation the male fetus exposes the gestating female fetus to his hormones. This causes the female fetus to emerge as a freemartin and to develop a masculinized brain. In some cases, the freemartin's reproductive tract is affected, and she winds up sterile. A cow freemartin can even take on XY chromosomes from her male twin, making her what is known as a chimera, although this phenomenon is relatively uncommon among sheep.

Gay rams also appear to be affected by hormone exposure during fetal development. During the process of gestation, hormonal exposure causes a part of the hypothalamus that controls the production of sex hormones to become enlarged in freemartins and reduced in rams. That seems to be the source for the altered sexual preferences. This same pattern in the hypothalamus has been noted in homosexual men. Scientists have found as well that gay men tend to have a "significantly thicker anterior cingulate cortex" and a larger "left occipito-temporal cortex"—a brain more like the one we would expect to find in a heterosexual woman.

So is this fetal hormone exposure the cause of homosexuality in humans? One theory that's been widely discussed has been put forward by Richard Wrangham. His belief is that homosexuality in the form seen among people is a side effect of selection for tameness. In his view, over the last hundred thousand years human societies have chosen to kill off the most dangerous and aggressive males, and this has produced a more docile man, one who, in some instances, will not fight over females because he is not sexually attracted to them.

There may be something to this idea, but it's worth considering a second, subtly different explanation. This is that the phenomenon of preferential male homosexuality among humans may be related not to the elimination of aggressive, heterosexual males from the gene pool but because the brain design called for by our evolution demands greater sociability. That is in line with the generally accepted evolutionary rule that homosexuality is most often observed among herding species: not only rams and sheep but giraffes and right whales. Thus, that humans display the impulse toward exclusive homosexuality is by itself a strong clue that we should best be understood as a herd animal and not as a primate.

This might not be a full explanation of the phenomenon either though. In humans it could be connected to the invention of spoken language. Exclusive male human homosexuality is reported to be roughly four times as common as exclusive lesbianism. Recall then that sufferers from Asperger's syndrome are overwhelmingly male, and the condition is associated with an exaggerated pattern of male brain development and lack of social awareness. Granted the necessity of speech and of understanding subtext for human survival, it may be that spoken language by its nature favors less masculine males and that there has been evolutionary pressure for a more feminized brain.

Let's return to the research that gay men tend to have a larger anterior cingulate gyrus and an expanded left occipito-temporal cortex. This part of the brain acts when we form words into lexical patterns. Hence, it plays a critical role in the process not only of empathetic feeling but speaking and reading. What we seem to be seeing among homosexual men are the characteristics of a brain that has been especially designed both for language acquisition and sophisticated social awareness. Could mild hormonal exposure upon a developing fetal brain be amplifying those qualities and male homosexuality be a side effect? While this is a possibility, the changes in gray matter volume associated with male homosexuality tend to turn up in other areas of the brain, including the prefrontal cortex and the precentral cortex.

Regardless, this is but one example of a larger point. As we learned how to talk, we changed the whole direction of our evolution. This was far more important and is more fundamental to what it is to be human than upright posture, hairlessness, or the opposable thumb. Apes can stand up on their hind legs, they can grasp objects with their hands, and, as we have observed, they can make tools. However, they have a limited ability to communicate with one another through vocalization. When chimps learn our words, they do so by recognizing gestures. Unlike dogs, they struggle to listen well enough to learn to understand vocalized statements or commands. Yet we know something about the other creatures that excel at this. They are herd or pack animals. Is it a coincidence that so many patterns of human action place us alongside the other large-brained herd and pack animals, most especially those gifted at communicating with one another?

In listing the mammalian orders in which homosexuality is common, I left out two. The first was the primates. I did so because homosexuality has a different function among primates than it does in humans. As we have said, chimp homosexuality is a means to prevent conflict. As Goodall has noted, that a male chimp grasps the genitals of another male is not an indicator of homosexual preference as it would be in humans.

Homosexuality is widely observed, too, in chiroptera, the bat species. Indeed, twenty-two bat species have been shown to engage in homosexual acts in the wild, and while bats are not herd animals they are notable for their complex means of communicating by vocalization. This includes the human and cetacean propensity for singing. Moreover, every single bat species that has been identified as engaging in homosexual activity in the wild also arranges itself into colonies. None is solitary. All this is leading us toward one point: Our heavy reliance upon vocal communication is not an incidental aspect of human existence, and it brings us back to our sociability.

Once humans developed speech, it was inevitable that we would domesticate ourselves. It was the principal cause of this, and it is both the reason that we can live peaceably among ourselves and the source of many of our problems. The requirements of a species with language are a higher degree of social engagement and listening skills than are found among apes. Neuroscientists and primatologists have long known that there is a relationship among the various primate species between the size of a species' brain and the size of its social grouping. This is to say that primates with bigger troops and bands have larger brains. It's been speculated that this may be because this requires more brain power available for knowledge of the different individuals in the troop. Think then about what the combined effect of our itinerant existence, our spoken languages, and our practice of identifying people by their names is. This allowed us to meet and remember many more people. It also meant that we *had to* recognize more faces, along with all the various people's passions, quirks, interests, and voices. This requires more brain power. But it is narrowly focused. It does not lead to enhanced skills at spatial awareness, agility, or tool-making. It calls specifically for more recall of individuals, and the expansion of our social web means that more of our activities, thoughts, and daily rituals become occupied with staying attuned to it.

To get a better sense for this, picture a middle-aged woman you know who has an accumulation of friends and acquaintances from high school, college, and various jobs. Since a count of those she knows might go well over a thousand, it is no longer proper to speak of her troop. Each day she joins up with her various cliques, and she tries to fit into them. But the life-long process of meeting so many people and interacting with them is part of something bigger. You cannot be influenced by one thousand other individuals and be truly independent. In the same way, when we enlist in the armed forces, enroll in a university, are hired by a large company, or join a union we are amalgamating ourselves with an enormous flock. We do this as well in identifying by nationality, ethnicity, religion, social class, and political affiliation. All these are herds, and, however much we may flatter ourselves by saying that *we* are thoroughly detached and free in our judgments, no one is.

This makes evolutionary sense. Research on traveling hunter-gatherers shows that in their travels they encounter many tribes. These contacts need not be hostile. They do require courtesy, attentiveness, and familiarity, and they further extend the requirement that we learn how to integrate ourselves within a wide matrix. Thus, possession of language by its nature forces us toward other evolutionary adaptations. It is not merely that this requires a larger brain. We cannot and do not exist within the narrow sphere of the chimp troop bounded by his small strip of land.

Language is a tool for thinking broadly and memorizing enormous amounts of information about great numbers of people, many of whom are no longer even alive. It compels us to do that. Thus, in ways that are coded in our genes and in ways that need not be it makes us something chimps and gorillas are not. We become like a stone in a river that has been polished down. We lose the jagged edges and rough points. This suggests that we have other characteristics of herd and pack animals. Among these is a degree of intellectual passivity. Like the polished stone, we become like the other stones below the passing stream. As we move about with the others in our troop, we are carried along by the flowing waters. We assist. We listen. We obey.

This is what herd animals do. Herd creatures adopt the outlook of those around them. They have a strong instinct to follow. As highly social and socially aware animals, we are obligated to. This is why humans are capable of joining cults and of killing ourselves or others on their behalf. It is why we will end our lives to advance the group, or, because we are capable of deep shame, we will commit suicide if we imagine that we have failed to live up to our sense of what we believe our friends and family expect of us.

Other proofs that our social nature requires us to conform? We build giant temples as a part of a collectivity in order to share in a common faith and devotion and to blend in with and placate those who are most devoted. At the same time, we are obsessed with status, and we wear clothes not merely to make ourselves sexually desirable but to indicate our identity within a group and our sophistication and rank. We decide matters by election, and we conclude that the opinion of the collective is a profound form of validation. We tell ourselves that things are true because others believe them. Or else we rely upon custom and heredity in determining our leaders, thereby showing an even greater esteem for the opinion of others, reverencing not only the views of the living but the dead.

For all this, language plays the greatest role not in our relationship to death but rather in the manner of our birth. Human babies are neonates. This would appear to be a redundancy as it literally means that our babies are newborns. It refers though to something that's critically important: For the first months of a child's life, it is completely unable to take care of itself. This contrasts with the readiness of nearly every other newborn animal. Baby colts and giraffes, for example, can walk right after they leave the womb, and newborn chimps and gorillas crawl along, however tardily.

Humans cannot because our brain is so big that if it were any larger it would be impossible for it to escape the birth canal. Consequently, we have to be born before we are ready to walk, and, as we all know, months of dependence follow. The gargantuan risk posed to women by bearing neonate children—babies who are so large with such big heads—increased the risk of maternal mortality. Analysis of past data shows that in the eighteenth century a typical woman in Europe faced a probability of dying of between five and twenty-nine times per one thousand live births. That means that a woman who had been through seven deliveries faced anywhere from one chance in thirty to one in eight of dying in the act of bearing a child. Combine that with the longer period needed to raise a human, and you have a situation in which women had to be able to trust those around them to raise their children since the likelihood that they would die before their children came of age was so great. This put a premium on cooperative amiability. This may partly explain why women tend to be better than men at reading facial expressions, and they display this superior skill at the earliest ages. Acute social awareness and the ability to read faces are also requirements for women as they offer an avenue for recognizing worrisome signs of sickness and ill health in children.

The survival of a parent is less important for chimps and gorillas than it is for humans or whales and dolphins. Studies have found that once he has reached age two a gorilla's likelihood of survival is not diminished if his mother dies. Chimps need their mothers far more than gorillas do, but they still require less instruction and for a shorter time than humans or cetaceans. The contrast between primates and large-brained herd animals in the process of development shows up before that though. It can be perceived through the length of time spent in the womb.

Elephants and giraffes have among the longest gestation periods of all animals: eighteen to twenty-two months in the first case, and fifteen months in the second. Likewise, dolphins and whales have notably protracted gestation periods. The shortest dolphin and whale gestation periods are about ten months, but orcas and many whales don't come out for fifteen to eighteen months. Other highly intelligent herd creatures follow this same pattern as is shown in camels (thirteen to fifteen months) and rhinos (fifteen to eighteen months). Also notable is the prolonged development in utero required for a number of the pinnipeds, like manatees (thirteen months) and walruses (fifteen to sixteen months). Chimps and gorillas have just eight-month gestation periods. How does that compare with us? Adjusting for our status as neonates, our effective gestation period is probably about twenty-one months. That's a little greater than an elephant or a giraffe but between two and three times as long as that of a chimp or a gorilla.

What explains this? The demands for our later intelligence and knowledge of speech mean that our brains must double in size during the first year of our life. Our survival depends upon this rapid post-partum expansion. That comes before everything else. During that year, 65 percent of our metabolism goes to brain development, and only 8 percent is directed toward the construction of our muscles. Remarkably, though, more than one-third of our brain capacity is still not present when we have reached our first birthday.

There's research on attempts to learn a language that suggests it requires thirty years to master one fully, and teachers who work with immigrant children have found that even the brightest children don't learn many aspects of a language for between five and seven years. Moreover, as we all know, small children trying to learn their native tongue are still working on the essentials behind concepts like irony and sarcasm when they are eight and nine years old. That they spend great amounts of time contemplating jokes like "Why did the chicken cross the road?" tells us how complicated this process is. Hence, much of the brain activity of the first year of life is focused on language acquisition. That may, in fact, be one of the reasons why children take so long to crawl and then walk. Our brains appear to be expending the bulk of their time and energy on something more difficult and more important to our eventual survival: learning to speak and to understand those around us. So it makes sense that the periods of gestation and rearing for elephants, cetaceans, and humans are longer than those for primates: All these species rely upon sophisticated versions of oral communication and the social relations arising from it that chimps and gorillas lack.

In our case then, we can begin to see how a mutually reinforcing pattern of effects was set off with spoken language's invention. As it forced women to bear neonate offspring, it set them at risk of death from bearing monster-sized babies. Then, as human brains got larger, female social relations had to change. Taking care of a baby that cannot take care of itself requires assistance. So women had to develop enhanced abilities to cooperate with other women. That was especially true because humans mature later. Altogether, this favored highly communal patterns of raising children and a greater amount of empathy and altruism among women. Where chimp mothers sometimes eat their children, human mothers teach their babies language so they can understand the metaphor of a parent consuming its child.

At the beginning of this book, I said that there were some iconoclastic primate researchers who had attacked the standard anthropological models of humans. Among the most influential is Sarah Blaffer Hrdy. Another was one of her college professors, the evolutionary biologist E. O. Wilson. These two influential academics probably did the most to undermine the belief that our species could be readily understood through observation of chimps and gorillas.

A two-time winner of the Pulitzer Prize and the recipient of dozens of other major awards, Wilson was the most eminent evolutionary biologist in the world at the time of his death in December 2021. Yet in his final years, he came to argue that the salient characteristic about humans was not our kinship with apes. He instead emphasized our capacity and instinct for cooperation.

Wilson's main area of study was ants, and he was universally acknowledged to be the world's foremost expert on them. He liked to point out that their ability to work together was at once an unusual trait within the animal kingdom and one that had made them almost uniquely successful. As their lifelong champion, Wilson liked to startle people with statistics on how abundant they are. Speaking in his mild Southern drawl, Wilson would astonish audiences with the raw numbers, informing his listeners that there are one thousand trillion ants in the world, and that the total weight of all "social" insects is 80 percent of the mass of all the bugs on earth. As he put it in one of his books, "Brazilian Amazon ants and termites together make up more than one-quarter of the [jungle's] biomass—which includes everything from very small worms and other invertebrates to the largest mammals. Ants alone weigh four times as much as the birds, amphibians, mammals and reptiles combined." Wilson recognized that this was a direct function of their ability to work together, and he provided hundreds of examples of this.

In 2010, Wilson made the connection to humans explicit by coauthoring a controversial article on altruism. It started out with a discussion of a surprising discovery: Some solitary bee species placed in lab settings naturally set to work with other bees. This is to say that they appear to be genetically inclined—programmed—to assist bees they are unrelated to. The article suggested that the rise of humans might be similar: Our ascent might be founded in a comparable and equally rare yet innate impulse toward empathetic collaboration.

This idea was not welcomed in academia. The source of the anger it provoked was especially to be seen coming from one clan: the primatologists. Assisting one another, showing a strong cooperative impulse, they attacked the belief that humans naturally form up into collaborative groups. The irony, it seems, was lost on them.

One who was conscious of this, however, was Wilson's former pupil. Like Wilson, Hrdy has made the human instincts toward cooperation and altruism one of her principal subjects, and a case can be made that she was the person who pushed Wilson toward the view that our capacity for unselfishness and collaboration strongly parallels that of other social animals and that it is the key factor explaining our species' rise. Yet in putting forward this idea Hrdy was making herself far more of a target for attacks from the primate research mainstream. For although she is now retired, her position at the

University of California, Davis, was as a professor of anthropology and primatology.

Hrdy's work has focused on the role that females have played in the process of primate evolution. Over the last four decades, this research completely overthrew the established thinking on the topic. It is the subject of a later chapter. What must be pointed out here is that Hrdy was among the first primate researchers to see that human sociability—or, as she terms it, our *eusocial* nature—sets us off from nearly all primates. As evidence of this, she pointed to a human behavior that is almost unknown among primates: nursing the offspring of other females.

Recently, Chinese primate researchers have found that female golden snub-nosed monkeys will sometimes wet-nurse for dominant females. Golden snub-nosed monkeys share one important trait with people as they are one of the few primate species living outside the tropics. Found in the highlands of rural China, they have to work together to keep warm and to find food during the cold months. But they are unlike humans in that they are highly promiscuous herbivores.

Which species are most inclined toward wet-nursing? As Hrdy has noted, this is characteristic of "canids, meerkats, and other cooperative breeders, or the sterile castes typical of eusocial breeders." The last category of creature is found among social insects: ants, wasps, and bees. It's also observed among those naked mole rats we mentioned at the beginning of this book, creatures that have a division of labor in their colonies. In other words, women will nurse babies for other women because our nature places us alongside cooperative hunting species, highly social rodent species, and bugs famed for their collaborative impulses—but not all but one of the primate species. Hrdy further observes that the habit of wet-nursing is so long-established among humans that it's referred to in Hammurabi's Code. That ancient set of laws listed the designated punishments for women employed as wet-nurses who were secretly providing their milk to more than one child as a way to earn extra money.

CHAPTER 7

HORMONES, HUMANS, HONEST SIGNALING, BIRDS, OCTOPUSES, AND CHIMPS

SO FAR WE HAVE MOSTLY been talking about how our minds are unlike those of a primate in terms of what has been learned through brain imaging. That leaves out a giant part of our story: biochemistry. This is the area of research that tells us about the neurotransmitters—the chemicals the brain uses to message itself—and our hormones. Hormones work in two ways: They direct many of our bodily processes, and they guide our basic drives, telling our brains what to think and feel. For this reason, they are critical to understanding why so much of our behavior is unlike apes.

Let's start with one of the most essential hormones: testosterone. While male chimps and gorillas do not have higher levels of testosterone than human males, female chimps and gorillas do have far higher levels than human females. Testosterone levels vary from 15 to 70 nanograms per deciliter in healthy premenopausal women. An average level in an older woman is 12 nanograms per deciliter. Female *Pan troglodyte* chimps have average levels of 180 nanograms per deciliter, and female bonobos have average levels of 170 nanograms per deciliter. So adult female chimps have from 2.6 to 12 times as much testosterone as young women, and 15 times as much as older women. These levels are consistent with the greater levels of violence and aggression displayed by female chimps.

These are not their only effects on the brain and behavior though. What we may be apt to ignore is that testosterone is the most important sex hormone. Bodybuilders who take anabolic steroid supplements in order to boost muscle mass say that it makes you supremely confident, bold, reckless, angry—and sex-obsessed. This is not a minor point. During the 1970s and early 1980s, when the gay liberation movement was taking place but before the AIDS epidemic had struck, it was reported that some male homosexuals had engaged in sex with hundreds of men. We know that bonobo females have sex with hundreds of males during the brief period when they are in heat. Should that be a surprise when they have testosterone levels that are so much greater than those of women and brains that are not fashioned for a reflective nature?

These facts also offer us hints about men and their behavior. When men spend large amounts of time around young children, their testosterone levels dip. This makes them less dangerous and aggressive, and it increases the survival chances of their offspring. That improves the chances that their genes will pass on to future generations. This is in line with what animal researchers refer to as the "challenge hypothesis." This theory holds that males of a species with high levels of aggression and competitiveness in mating caused by elevated testosterone levels tend to be uninterested in parenting. However, males of a species that are cooperative and less competitive regarding mating participate in raising the young and thereby assist in the passing on of their genes. The challenge hypothesis was developed through study of mating among birds, but it has proven to be predictive among mammals as well. One of its corollaries is that the males of species that mate seasonally (e.g., groundhogs, minks) are more likely to be aggressive in intra-sex competition, to have high testosterone levels, and to be absent fathers.

Though that is not the cause for intra-sex competition between chimps and gorillas as they do not live at seasonal latitudes, there is abundant evidence that the general principles of the challenge hypothesis apply to apes—and humans. Research shows that human males exposed to higher testosterone levels in utero are significantly more inclined to be unfaithful to their wives and girlfriends, and men with lower levels seem to be more nurturing. It's additionally been proven that the amount of testosterone a fetus is exposed to in the womb can determine its sexuality and its future constancy. It even plays a role in how long our fingers are. High testosterone primes the body for more prominent ring fingers. Low testosterone leads to longer index fingers. Simply put, hormones influence many aspects of our bodies and our thinking. So that chimp females have between 2.6 to 12 times as much as young women and 15 times as much as older women isn't an incidental point. It guides the construction of our bones and the wiring of our brains.

All this relates to that second great wave of human evolution. When people became interested in making art around 75,000 years ago, our faces began to change. A Duke University study of human skulls suggests that shifts in facial appearance indicate a persistent trend "toward a reduction in the brow ridge and a shortening of the upper face, traits which generally reflect a reduction in the action of testosterone." They noted that *Pan troglodyte* chimps have more pronounced brow ridges, and that these appear to be caused by dramatic increases in testosterone during a male chimp's adolescence. This pattern is less apparent among bonobos. However, male bonobos undergo rapid ascent in cortisol levels in their teen years, and since this stress hormone induces aggression, the significance of the relatively smaller rise in testosterone may not be so great. And studies show that male bonobo testosterone levels scale up quickly when they are in the presence of ovulating females.

Variation in the serum levels of testosterone among different species has other implications. Testosterone spurs not only swagger but the confidence that comes with it. High testosterone levels improve spatial abilities, and chimps do well on tests of spatial ability. But testosterone also has adverse effects on intelligence. That was shown in a study by researchers at CalTech and Wharton. They found that when people were given a gel containing testosterone they were quicker and more impulsive in answering simple math questions. This caused them to give back more wrong answers. That suggests that chimp females are not only more aggressive but less able to reflect and to act as sources of corrective guidance, whether to temperamental males or for children in their midst.

Testosterone is not the only sex hormone, of course. Almost equally important for women is oxytocin, the "love" hormone. This chemical makes us yearn for physical intimacy, and it rises dramatically at the moment of the birth of a child and during breastfeeding. Women and men also feel its effects during lovemaking. Inevitably then, it plays a critical role in inspiring a mother's love of her children and in pair-bonding between men and women.

After women fall in love and become sexually attached, their oxytocin levels go up for approximately six months. It happens that a fertile woman regularly having sex has a 30 percent chance of getting pregnant in the first month of a relationship and a 75 percent chance of conceiving through half a year, and, if she has not done so through six months of regular intercourse, there may be something wrong with the man's potency or the couple's compatibility. It would make sense at that point for her sexual interest to wane, and many men and women have noticed that this length of time also corresponds to the period when women are most interested in sex—often more so than their mates. Isn't that logical? And doesn't it follow that during this stretch of time women desire lovemaking as much for the feeling of attachment as sheer lust?

Yet, as the manner in which hormones act can be different in different species, a group of primatologists working in the Ivory Coast came up with a seemingly oddball theory about oxytocin. Their idea was that it might be drawing a chimp troop together before it fought with another troop. To see if that was so they used high-powered cameras that allowed them to keep an eye on the chimps from a safe distance. Then the researchers went and checked the urine the chimps deposited in the soil, looking for changes in oxytocin levels. And sure enough they found that before fights oxytocin levels actually doubled, rising in both male and female chimps.

The spur to the investigation was a study done with humans. It showed that when groups of humans were given nasal sprays of oxytocin that they became more effective at working together in competitive interactions with other groups. However, in humans while it made small groups more cohesive, it did not make people more aggressive. There was no increase in the instinct for attack. In fact, it reduced this, even as it bonded people who saw themselves facing competition. It's possible that it's functioning in the same way in chimps. But the more obvious conclusion seems to be that although it pulls humans together and reduces our aggressiveness that it draws chimps together and increases theirs.

There is also strong evidence that testosterone and oxytocin have different influences on our capacity to trust one another. When we are exposed to oxytocin and then shown pictures of other people's faces, we are apt to say that they are trustworthy. But testosterone has exactly the opposite effect. This is likely a crucial part of the reason why women can and do raise children together, but chimp females can't and don't.

That shifts in hormone levels have played a part in our domestication is shown in our body chemistry in another way. It turns out that women have much higher levels of a paired set of steroid precursors known as dehydroepiandrosterone (DHEA) and dehydroepiandrosterone sulfate (DHEAS) than chimp females. On average, women have three times as much. These hormones are used to make estrogen and testosterone, yet they have effects that reduce our impulses toward aggression. They do this by acting to control the cortisol levels in the body. That serves both to quiet our feelings of anxiety and to diminish the violent urges that arise from them. It has also been demonstrated that DHEA can be a treatment for mild depression. Perhaps most significantly, research reveals that DHEA plays a critical role in childhood brain development. But, while it is one of the most abundant hormones in humans, it is not even found in most primates after birth.

However, it isn't only a matter of how our hormones are distinct from apes. We have a different mix of neurotransmitters in our brains than chimps and gorillas. Neurotransmitters are the chemicals that our nerves use to send messages to other nerves. Most nerve cells are designed to put out just one of these. Neuroscientists refer to this as Dale's law in honor of the scientist who first observed the principle.

Although more than sixty have been identified, just a few are responsible for the greater amount of activity taking place in our heads. These main neurotransmitters are serotonin, norepinephrine, dopamine, acetylcholine, and gamma-aminobutyric acid (GABA). The last of these is involved in sleep, and it helps control dopamine. That's the neurotransmitter that makes us feel pleasure and love.

In an earlier chapter I mentioned HAR1, the human accelerated region gene. One puzzle that has long perplexed neuroscientists is why only humans appear to suffer from schizophrenia. That riddle prompted a doctor at Mount Sinai Medical Center in New York to see if the HAR1 gene lies close to the genes associated with schizophrenia. Intriguingly, it turns out that it does. That follows logically as HAR1 plays a role in GABA production, and schizophrenia occurs when there is a lack of GABA. This affects activity in a part of our brain called the hippocampus, a region that deals with dreams and the setting down of memories. When the hippocampus is absent of sufficient quantities of GABA, dopamine production becomes excessive. That causes the brain to stop being able to distinguish fantasy from reality, leading to hallucinations and delusions. This particularly affects humans because in the parts of our brains that provide us with the emotional stimuli that reward us for our actions we have more dopamine and less acetylcholine than chimps and gorillas do. That makes sense as humans need to feel the emotions of love and connection that dopamine codes for, and we have a far greater need for our imagination and our powers of recollection. By contrast, apes require the instincts toward fighting and aggression that high levels of acetylcholine awaken. And the differences aren't small. One study that looked at the production of these chemicals found that chimps had two to three times as much acetylcholine as humans.

Healthy humans have relatively consistent quantities of GABA throughout life. The one period when much change shows up is in our teenage years. At that point GABA levels will rise in the hippo-campus. Schizophrenia is apt to follow when it declines instead. On the other hand, when GABA levels are excessive in early childhood, this can induce autism. In effect, if the hippocampus is too controlled, the child becomes unable to read subtext.

Seemingly, the boost in GABA during a boy's adolescence hap-pens because evolution is preparing human males for the tensions and pressures of young adulthood. It's doing this by providing them with more of a neurotransmitter that will offer them a degree of self-con-trol. This will partly counteract their soaring testosterone levels.

At present there isn't any data on GABA levels in chimps and gorillas. But there is some on rhesus monkeys, and that fits in with the picture that we see in apes. When rhesus monkeys reach adolescence, their GABA levels abruptly decline. The falloff is steep. At one point, it drops by more than half in a span of a little more than a year, and it continues falling through the whole of their entry into adulthood. In effect, it seems that when rhesus monkeys reach this age they are supposed to possess less of the neurotransmitter that helps them put a rein on their emotions. When they grow up, it appears that nature is designing them so that they will be wilder. For rhesus monkeys, "maturity" is attained through its absence.

Nor is this the only brain chemical that follows this pattern in which humans have been programmed for self-control while apes and many monkeys are steered toward conflict and aggression. Geneticists have known for some time that a protein involved in the development of skin cells also affects the amount of impulsiveness we feel. Called epidermal growth factor (EGF), high levels of EGF cause people to be "novelty seekers." They exhibit traits like extrava-gance, disorderliness, and excitability. By contrast, those with low levels of EGF are apt to prepare better for future events, to display shyness, and to worry about what's to come. Geneticists had assumed that the patterns of EGF manufacture would be alike among primates, but when they tested for this they found that humans have distinct chemical pathways that deter much of its production. Moreover, it turns out that convergent evolution has led to vital instruction sets in humans that are alike to those found in dogs but different from those in chimps. Some of our instincts for self-control and thinking ahead, it seems, are alike to those of our most beloved pets yet quite unlike those in chimps, baboons, macaques, and rhesus monkeys.

The differences between chimps and humans seem especially to apply to men. It may, in fact, be said that maleness means something different for humans than it does for chimps. Geneticists have demonstrated this in another way. That's through a family of genes at the site of Theophilus Painter's old area of study: the Y chromosome. The whole grouping of genes on the Y chromosome responsible for sex determination is different in human and chimpanzee males. As these genes have shifted little over hundreds of millions of years among most animals, researchers assumed that they would be nearly identical in humans and chimps. But they actually found "extraordinary divergence." In their words, "[T]hey differ radically in sequence structure and gene content, implying rapid evolution during the past 6 million years." Since we split off from chimps, it turns out that fully 30 percent of the genes coding for maleness changed. They have transformed to the point that they are no longer homologous—alike to one another.

I mentioned at the beginning of this book that my interest in neuroscience commenced as I learned about my father's research when I was a child. Before he became a university professor, he was employed by the company that would eventually become the pharmaceutical giant Bristol Myers Squibb. Working as a researcher tasked with developing antianxiety drugs, his focus was upon a family of medications known as benzodiazepines. Most of us know them by their trade names. Among these are Valium, Klonopin, and Xanax.

As my father did not like the corporate world, he fled from it and went into academia. Nonetheless, he remained curious about these drugs, and with the backing of his university he decided to investigate their effects on laboratory animals. His findings were not what he expected. To his amazement he discovered that these "minor tranquilizers" had an opposite effect on rats as they did on humans. While they induced calm and sleepiness in humans, they made rats aggressive and eager to kill mice. Perhaps even more surprisingly, he found that amphetamines seemed to pacify the rats, making them less inclined to attack. The lesson he took from this was that it was impossible to understand human reactions to drugs exclusively through models that were based on animals that lacked complex patterns of thought and language: *human patterns*. This struck him so strongly that he stopped using lab animals for research, shifting the whole direction of his career. This is worth bearing in mind.

Regardless, research on the effects of hormones *on humans* does tell us something about how they affect us, and study of oxytocin offers us a partial answer to the question of why it is that firefighters and soldiers will risk their lives as a matter of occupation, assuming a profound obligation on behalf of people they have never met. Psychologists suggest that this kind of valor is founded in a love for those with whom the soldiers and firefighters serve, and this fits with what we have come to know of oxytocin. It's also aligned with the idea that some of the appeal of being in an army may come from the higher levels that camaraderie produces and that the suicidal despair veterans sometimes suffer on returning home may be linked to a subsequent drop-off in oxytocin levels.

That gets us to another point. There is a difference in what a human hopes to gain from a dangerous situation and what a chimp seeks. A primatologist will tell you that a chimp shows fearlessness in combat because he believes that this may lead him to raised standing within the troop. Whether his aim is enhanced reproductive odds through higher status or the pride attached to this, his desires are specific and practical. In de Waal's words, "Chimps never make an uncalculated move."

Humans can be inspired by purely abstract goals and ambitions. Notions of honor extend beyond our own term of life, while in a religious war we may be thinking about another one. Thus, we see once again that among humans intangible aspirations can outweigh material goals and self-interest. We can conceive of ourselves standing on a dais, receiving a medal for our steadfastness and courage. We can dream of others speaking respectfully of us after we have passed away. We can become attached to an ideology or a faith and be willing to sacrifice ourselves for it.

The idea of profiting in another life and the human capacity for altruism find their union in the phenomenon of priestly celibacy. A monk's cowl and a nun's habit are in some measure soldiers' uniforms, ones that join together two traits notably deficient in chimps and gorillas if perpetually evident in humans. Monks and nuns are but two of the many groups of people devoted to helping others. Their existence points to another problem with the primate model of human behavior. Working from it, you will inevitably find it difficult to understand why people set out with the somewhat analogous goals of becoming nurses, social workers, veterinarians, and teachers. These wishes cannot be ascribed solely to self-interest. Nor are these alike to the aspirations or plans of any chimp or gorilla. It is only by rejecting these simplistic models of human behavior that whole categories of human action start to make sense.

The concern with a future life displays itself in our practice of burial. There is evidence of human death rituals dating back to the Middle Pleistocene: 92,000 years ago. That was revealed in the 1960s when excavators at the Qafzeh Cave near the ancient town of Nazareth discovered the graves of fifteen people. Alongside them were ochre-stained tools, deer antlers, and shells that appear to have been used in a ceremony associated with their placement in the cave.

Chimp behavior is different. As we have noted, chimps grieve when those dear to them die, and they have recognizable practices with regard to corpses. Chimp mothers will hold on to the body of a dead baby for weeks as though testing it for signs of life, and such behavior has been observed in other primates. But far more alike to the actions of people are those of dolphins and elephants. Like humans, they collectively show their respect by making group passages about their kinsman's dead body. They will stand watch upon it, preserving it from attacks. In the case of elephants, this will be to protect it from jackals and vultures. Martin Meredith, a famed elephant researcher, observed that on one occasion:

> The entire family of a dead matriarch, including her young calf, were all gently touching her body with their trunks, trying to lift her. The elephant herd were all rumbling loudly. The calf was observed to be weeping and made sounds that sounded like a scream, but then the entire herd fell incredibly silent. They then began to throw leaves and dirt over the body and broke off tree branches to cover her. They spent the next two days quietly standing over her body. They sometimes had to leave to get water or food, but they would always return.

There are actually reports of elephants that trampled humans burying the people they had killed. Dolphins will group together around a corpse to ward off human swimmers. Likewise, there are accounts from all over the globe of corvids meeting up in improvised acts of mourning.

The cases of filial cannibalism among chimps might incline us to think that chimp mothers are uncaring about their offspring. In fact, to a very great degree a chimp's idea of love can be equated with a mother's tender feelings and concern for her children. Insofar as we can speak of chimp love as something akin to human love, it is most clearly displayed in maternal affection, and accounts of the sadness and despair felt by chimp mothers for their dead offspring are harrowing. As we have mentioned, this intense devotion is especially manifested between mothers and their sons. Chimps are not typically weaned until they are five years old, and chimp males develop keen lifelong attachments to their mothers. Consequently, it is chimp females who are likely to leave the troop, not chimp males. All this must ring a chord in our hearts and make us conscious that chimps are relatively near to us genetically.

One must ask though: In what way does our knowledge of this much advance our understanding of ourselves? Were we previously unaware that mothers tend to love children? The structure and design of the human brain isn't only alike to the ones possessed by chimps and gorillas. The three-part arrangement—brain stem, cerebellum, and cerebrum—is common among mammals, and in many, if not most, mammal species mother love is an essential part of life and the life cycle. What's more, the way in which maternal affection manifests itself in chimp relations is different from what we see in humans. To take but one example, when a young male chimp loses a fight with an older rival his normal way of responding is to go to his mother and seek her tender embrace as consolation and for support. Wouldn't a seventeen-year-old boy be a bit uncomfortable and ashamed to behave this way? As we have mentioned already—and all know—human teenagers are looking for encouragement from their peers and from incorporation into their various rites and rituals. That is where our adult identity emerges as it is where we engage in youthful bonding. It is where we acquire our taste in music and clothes. Of all the herds to which we belong it may be the most intrinsic.

Advocates for chimps and their "humanity" will often in engage in special pleading, trying to persuade listeners that chimps are empathetic and caring. Thus, Frans de Waal wrote in one of his books of an incident in which a chimp leaped into a moat to help another chimp that had fallen in. But it's an anecdote—no more. Moats are placed around chimp enclaves because they are so dangerous and uncaring that we are afraid to be near them, and, by and large, chimp life is a zero-sum game. There are many reports of humans jumping into moats to save chimps, but only one has ever been recorded of a chimp trying to save another chimp from drowning.

Why is this? Why is it that we take such pride in contributing to hospitals and helping orphans? The term "altruism" is a recent introduction to our language. It comes from French. The man who invented it, the philosopher Auguste Comte, intended it as an expression that would represent the opposite of egotism. The best examples of it to be found among other vertebrates come from whales, dolphins, rats, dogs, and wolves. Each returns us to that fundamental question in evolutionary biology: Why do creatures behave altruistically?

The general explanation within the field is something called the green beard effect. This says that a trait might survive if all the animals that possessed it gave preference to others of their species that had it as well. The term comes from the idea that a creature with a green beard might favor others with the same attribute, even though the trait did not by itself offer a survival advantage to him.

With this in mind, let's look again at our list of mammals known for altruism. What they have in common is that they are either highly intelligent hunters who must work collaboratively or potential prey that are not notably fast or able to rely upon the protection of a herd. The first category includes whales, dolphins, dogs, wolves, and humans. The second takes in rats and some species of bats.

We have already spoken of the unique evolutionary problems faced by cetaceans. This would explain why dolphins are especially known for their altruistic deeds. Cetacean researchers have noticed that dolphins will aid injured companions by pushing them to the surface so they can swallow air. In other cases, dolphins have helped beached whales get back into the ocean, and they have formed circles to protect people from sharks. These actions placed the dolphins at risk for their own lives. They are behaving as we do when we put ourselves in jeopardy, whether to save human lives or those of other animals.

Rats face special difficulties too. As they are not herd animals, they cannot group together and ward off predators with the danger of a stampede. Nor do they have a porcupine's quills, a snake's poisonous bite, a chameleon's gift for protective coloration, or a skunk's attack of stench. Small and weak yet continually exposed, they must learn to work together to ward off strikes by larger predators. The habit of helping even unrelated rats is their green beard, and it has undoubt-edly played a role in their survival.

In laboratory experiments in which rats had a choice between eat-ing chocolate cookies by themselves or opening a pen holding other rats and sharing the cookies, rats elected to assist other rats and give up a portion of the cookie haul. Rats can smell hunger in other rats, and when they do they are more likely to give food to those that are famished and in need. This contrasts with the behavior of chimps. On occasion, they will "pant-hoot," alerting others to the presence of food, but they do this when their potential meal companion is power-ful and of use to them, not when they are low-ranking and ravenous.

Proof that humans can be profoundly altruistic can be found in Keithville, a small town in Louisiana. To get to it you head east from Shreveport. After about twenty miles of driving, you switch onto state highway 171. Take it south and then circle back to the west, and you will spy a road called Chimp Place. It leads to a 200-acre property called Chimp Haven. Home to more than 300 chimps, it's one of eleven sanctuaries set up in the United States for retired apes. Another 200 are in Morgantown, Georgia, and other sanctuaries have been set up in Texas, Florida, California, and Washington. At each, a staff of caretakers provides meals to the chimps along with regular health checks.

Most of these sanctuaries were set up in response to a federal law, passed in 2002, known as the CHIMP Act. The acronym stands for the Chimpanzee Health Improvement, Maintenance, and Protection Act. Under the law, the United States government is obligated to "pro-vide for the lifetime care of chimpanzees that have been used, or were bred or purchased for use in research conducted or supported by" the federal government. Some who deal with the chimps on a day-to-day basis refer to it as chimp "social security." The law prohibits euthanasia of chimps used in research. There are also an additional twenty-three chimp and monkey sanctuaries in thirteen different African countries. The best-known in the United States is Chimp

Eden, while the Jane Goodall Institute's 1,000-acre chimp sanctuary is in South Africa.

It is not unusual for a species to have symbiotic relationships with other species. For example, cows and bulls rely upon egrets to pick out insects that lodge themselves onto their skin. Yet, with that exception, most animals are unconcerned with the suffering of other species, and for a long time humans were blithely indifferent to chimp misfortune. Worse, we were a cause of it. But, as the idea that we are like chimps grew in the public consciousness, the CHIMP Act advanced in Congress, and eventually it became law. As this was happening, major corporations and philanthropists were constructing reserves to protect apes and monkeys across Africa.

Here again we see the contrast between humans and great apes. While Goodall and gorilla researcher Dian Fossey set themselves in frequent danger to help the apes they studied, chimps and gorillas won't do the same for us. Obviously, one might argue that this reflects their poorer level of consciousness. But, even if this is true, it does not change the fact that the corresponding actions reflect differences in nature.

Anthropologists and psychologists influenced by the primate model of human consciousness have gone to enormous lengths to identify instances of altruism among chimps and gorillas in order to use this as an explanation for the acts of bravery and sacrifice that humans display. Yet finding similar demonstrations of unselfishness among great apes is difficult. This is so much the case that in his book *Our Inner Ape: A Leading Primatologist Explains Why We Are Who We Are* Frans de Waal commenced his account claiming that chimps are altruistic in the manner of humans by citing an example of *canine* altruism. Such instances, needless to say, are easier to reference.

This does not explain though why humans set up soup kitchens. We have already mentioned one reason: our status as cooperative hunters. There is an additional explanation we must return to that may be equally important: women cooperating to raise children, what anthropologists refer to as alloparenting, or cooperative breeding. This has been one of the particular areas of Sarah Blaffer Hrdy's interest, and through it she has largely destroyed the consensus view that humans can be understood as "another primate."

To grasp Hrdy's work and her importance it's necessary to know something of her background. Now in her late seventies, Hrdy was an intellectually precocious teenager. When she arrived at Wellesley in 1962, she was just sixteen. Two classes ahead of future secretary of state Hillary Rodham Clinton, she had matriculated at one of the country's most famous women's colleges just as second wave feminism was starting. The movement was part of a broader trend by which young people were questioning established thinking. After transferring to Radcliffe, Hrdy decided to major in anthropology. Her undergraduate field research was on a tribe of native Mayan-language speakers in Chiapas, Mexico. The prompt for her interest in primates lay in an article by the Harvard primatologist Irven DeVore. He was the first anthropologist to show that male langur monkeys frequently killed off the babies in their troops. Influenced by the "population explosion" theorist Paul Ehrlich, DeVore had come to the conclusion that this was a consequence of growing population pressures.

As a graduate student, Hrdy went off to India to test the theory. She came back to Cambridge with a very different idea: Male langur monkeys weren't killing off babies because of a fear of overcrowding but in order to advance their own evolutionary chances. That got her thinking about how the female monkeys might be responding to this and how they might be trying to fool the males. From this came her influential book, *The Langurs of Abu*.

Hrdy believes that the source of human altruism is our practice of alloparenting. Gorillas do engage in some alloparenting. But some chimps appear to receive none whatever. As a team of researchers recently concluded, "[chimp] babysitting may be a behaviour that rarely occurs." However, the difference between how chimps and gorillas rear their young may not in actuality be so great. For, as another paper on observation of interactions between gorillas and their mothers, notes:

[A]part from mothers encouraging their infants to walk within their first few months, the gorilla infants were more active in their own learning, rather than their mothers pushing them. It was the infants that took an interest in the activities of their mothers, resulting in the infants repeating their mother's behaviors. In chimpanzee and gorilla species, there are few reportings of active involvement of learning from mother to infants [which] has been recorded, where mothers have taken inedible items out of their infants' mouths, or administering active teaching.

These differences in rearing extend to feeding. Chimp mothers rarely provide their babies with anything more than breastmilk. By contrast, not only do human mothers go to great lengths to provide their children with as much food as they can, but anthropologists have learned that in hunter-gatherer tribes grandmothers will increase the amount of time they spend in foraging whenever there are grandchildren. We teach and feed our young on an intergenerational basis. For the most part, chimps and gorillas do not.

Hrdy's belief in the central importance of alloparenting as a cause of our evolutionary change prompted her to work on an experiment that provided intriguing evidence on behalf of her hypothesis. In 2014, she and a group of her colleagues published research on a test they had devised. It required the construction of a special instrument, a machine that permitted animals and humans to give a favored treat to someone else but not to themselves. Using human children and adults of fourteen primate species as their subjects, they demonstrated that the species that consistently displayed an instinct toward altruism were those that engaged in alloparenting, not those that hunted. This was so true that there was an almost linear relationship. Marmosets and tamarins and, most of all, humans were altruists. Chimps were not.

The problem with this research, of course, is that, as we have noted, chimps actually don't hunt much or all that well. Hence, while this provides compelling evidence that the practices of women in working together is a central part of the story of how humans evolved toward a capacity for collaboration and to mutuality, it does not prove that the human switch toward meat-eating and hunting didn't also play a part in the changeover. Regardless, we can say with confidence that among most species the competition is so keen that a tendency to assist unrelated kin provides no evolutionary advantage. Thus, among chimps, there is no green beard.

In any event, there is even another explanation for altruism. It lies in a concept called indirect reciprocity. This is the idea that each of us develops a reputation, and that altruism is a series of judgments about what we can safely get away with without incurring injury toward our reputation. Along with this, we make calculations about whether or not we believe that a particular individual would ever repay a debt. Over time a sociopathic personality will tend to be denied this benefit as his reputation comes to match his character. A species inclined toward selfishness will be largely without this form of altruism since overall levels of trust and expectations of reciprocity are diminished.

That difference in group behavior is further enhanced by a concept called Hamilton's rule. It states that when a creature knows that he is acting on behalf of kinsmen, he will act more on behalf of his relations proportionate to the degree that he knows they are related and that a likelihood of passing on their genes is enhanced. Hamilton's rule does not absolutely require that creatures know they are related for it to operate, but this greatly boosts its power. Obviously, this has implications in terms of the development not only of human societies but in any comparison of humans with chimps. That's because chimps never know their paternity.

There can be no disputing the claim that apes are more social than the mammalian species inclined to solitary hunting, like tigers and grizzly bears. But insofar as use of the term implies a high degree of awareness and concern for the interests of others in their tribe or family then we cannot rank them terribly high. Nor do they show much capacity for guilt or shame.

That's indicated by the fact that it's relatively difficult to find notable examples among chimps of what is known as "costly signaling." This is the phenomenon by which "animals (including humans) may send honest signals about desirable personal characteristics and access to resources through costly biological displays, altruism, or other behaviors that would be hard to fake." Humans engage in costly signaling in many ways. One of the most famous examples arises through what is known as conspicuous consumption: the purchase and display of expensive items that we buy *because* they are costly and superfluous. Economists explain the idea through the example of fancy watches. Those priced for a few hundred dollars are less appealing to consumers than similar ones costing thousands of dollars. A bond trader is eager to display his Rolex expressly because he had to shell out several thousand dollars for it, *and* it keeps no better time than his phone.

Amotz Zahavi was the evolutionary biologist who first developed the theory of costly signaling as a concept that applied throughout the animal kingdom. He saw it as the explanation for the development of unwieldy but attractive secondary sexual characteristics. The best known of these is the peacock's feathers. While the plumage appeals to peahens, it weighs the male down, making him less agile, and it requires a huge expenditure of energy to produce. The large antlers on a male deer are another familiar instance of costly signaling. They are proofs that the male is healthy, impractical as they are.

As Zahavi was an Israeli, he took a particular interest in one of the country's native birds, the Arabian babbler. We mentioned the babbler before as it's believed to be the only species other than humans that mates in private. This is proof of how social it is. Zahavi and his colleagues learned that a favorite activity of babblers is to take time out from foraging in order to go up to a high spot where they can perch and spot potential predators. Eyeing a foe, they give out a cry, warning their fellow birds of danger. The protective signal is provided to relatives and nonrelatives alike. So babblers don't do it simply to protect their kin as a means to help them pass on their common genes. Still, the babblers are most inclined to act as sentinels when they are around members of their group or flock. In other words, they aren't doing it completely unselfishly. They are trying to win points by showing other birds that they are responsible members of the flock. Researchers found that sentinel behavior is most common among creatures that co-breed, mutually rearing their children as people do. As such, other animals that are especially known for sentinel activity are pied babblers, jays, meerkats, and mongooses. It's telling then that chimps don't appear to do much costly signaling, and they don't have the exaggerated secondary sexual traits of social bird species like peacocks or herd animals like deer. Aside from their genitalia, male and female chimps are hard to tell apart. As will be discussed in a later chapter, this contrasts with humans who have a number of obvious secondary sexual characteristics, including male beards and women's breasts.

Even so, a part of our nature is alike to that of chimps and gorillas. That is undeniable. But we seem to have changed in essential ways that make us more similar to highly intelligent herd animals like dolphins or elephants than we are to chimps. In some ways, too, we may be more like cows and bulls. A friend who grew up on a dairy farm in Vermont once told me that when she returned home after a number of years away the cows immediately recognized her and came over to nuzzle their heads against her. This is not an unusual experience, and it points up the fact that while we have bred cows to be exceedingly docile and uncritical they have a keen emotional awareness along with a need for physical intimacy. These traits are alike to qualities found in humans. This gregarious warmth stands in opposition to how chimps most often respond to other chimps—or people.

It also contrasts chimps with dogs. I mentioned Emory University neuroscientist Gregory Berns earlier and his work on the caudate nucleus. In a separate study, Berns looked at whether dogs responded more to the praise they received from people they cared about or the food the people brought. Brain-imaging tests showed that in many instances they really did care more about accolades from people they loved than food. *That* is sociability. But it's unimaginable with chimps since their focus is not upon winning favor but in obtaining power and resources. In fact, even the tests that Berns and his colleagues performed would have been impossible with chimps as they required the dogs to undergo months of training so they could lie still for hours of testing by advanced brain-imaging machines. That the dogs were able to do this is yet another proof of their attentiveness and their likeness to us.

Think about that in light of another of the supposed proofs offered by primatologists in order to show that apes and monkeys are highly social: the amount of time that they spend grooming members of their own sex. A few points ought to be made with respect to this. First, it has recently been shown that the manner in which apes groom one another is based upon the way they were taught to do this by their mothers. They do not display styles of grooming that they adopt or learn by imitating their peers. This differs from humans, who, as we know, tend to pick up styles of dress, mannerism, and speech from their age cohort, and are intensely concerned with being part of the group.

Researchers have found that intra-sex grooming among animals is of two types. One is called allogrooming. This is when creatures try to help one another by picking out parasites they cannot reach by themselves. This kind of grooming is observed among herd animals like cows. In expectation of receiving grooming in return from another chimp, chimps will groom one another. Grooming also plays a role in the creation of alliances within the chimp troop. That function ties in within a second type of grooming, which is frequently observed among chimps, bats, and insects. It is one of the most important activities among apes, and it is profoundly social. There is no doubt but that these creatures groom one another to strengthen their social bonds: their feelings of attachment and connection. There is another overriding purpose though. The grooming serves to establish hierarchy and to reduce conflict and tensions within the troop.

But it is not like the behavior of humans. Ask yourself: When have you observed humans grooming others of their own sex? The only instances you are likely to see are at a hair salon, by lovers, or by women who are friends. It is not meant to show who is in charge or to break up fights. It is either a paid service or a display of affection.

Primate anthropologists might respond here by saying that an office worker will go into his boss's office and flatter him and that this is a comparable form of grooming. Yet it really isn't. Grooming among apes has been shown to release oxytocin, the hormone that we associate with bonding between mates and between mothers and their children. It is a palpable act with physiological results, and most often it appears to be unlike ours. We can achieve this effect by hugging our spouses and by speaking words of love and adoration to them. This is not how chimps behave as they don't have sustained connection or much intimacy with their mates. Instead, among apes the primary function of grooming is to reduce tension on an intra-sex basis within the context of an existing dominance hierarchy. Humans are tame, and, as emotionally and socially aware as we are, we would find the act of grooming another member of our sex as a means of demonstrating our obeisance to them to be shameful. Only when we are being compensated financially or under extreme conditions will we permit ourselves to be forced into such a humiliating act. We will massage another person as a means of acknowledging their power when we are placed in prison. Typically, however, we do it as an act of love. By contrast, chimps decide which chimps to groom based on the other chimp's perceived authority and status. In this, they are making a frank acknowledgment of their subordination. Simply put, humans display the grooming habits of herd animals, not apes.

Likewise, while chimps may groom members of the opposite sex as a means of seduction, this has no other purpose. No pair-bond is being created as they do not develop romantic attachments or connections to their lovers. Everyone who works regularly with chimps is keenly aware of this. In Goodall's words, "I cannot conceive of chimpanzees developing emotion, one for another, comparable in any way, to the tenderness, protectiveness, tolerance, and spiritual exhilaration that are the hallmarks of human love in its truest and deepest sense. Chimpanzees usually show a lack of consideration for each other's feelings which in some ways may represent the deepest part of the gulf between us and them. For the male and female chimpanzee there can be no exquisite awareness of each other's body—let alone each other's mind. The most the female can expect is a brief courtship display, a sexual contact lasting at most half a minute, and sometimes, a session of social grooming afterward. Not for them the romance, the mystery, the boundless joys of human love."

At the same time, it must be noted that many primates *are* alike to humans in how they mate and groom one another. For example, titi monkeys are highly monogamous, and most monkeys groom one another affectionately and not simply as a means of establishing dominance hierarchies or in order to avoid conflict. Yet our supposedly most alike relatives—chimps—are not monogamous, and their grooming is unlike ours.

Bearing this in mind, allow me to cite a few more examples of how humans are innately different from chimps and gorillas. First, think of what happens when you arrest a human and place him in a jail cell. Right afterwards you must keep him under observation. Why? Not infrequently the shame and fear induced by the experience prompt people to take their cell's bedsheets and hang themselves. This happens all over the world. It occurs even when people face minor charges. Apt to be aware that they will soon be arraigned and released, the overwhelming majority of those who kill themselves were not suicidal before their arrest and were not in ill health. Rather, the sense of terror, failure, and the worry about what others will think of them afterwards cause them to commit suicide. Similarly, teenagers—those with nearly their whole lives ahead of them—will sometimes slit their wrists or swallow bottles of pills following inconsequential failures. While this reflects their adolescent lack of judgment and perspective, we must ask ourselves why such behavior is common among humans but unimaginable in apes. To the same degree, what does it tell us that young women will stop eating because they have developed a fear that they are obese, but anorexia has never been observed in chimps? All these actions are proofs of how our identity is bound up with concerns about how we are perceived by others.

Embarrassment and disgrace are human concepts in a way that they could never be with apes. Shame is an emotion that arises in animals that are deeply concerned with how those around them think of them. As the example of dogs shows, this has relatively little to do with brain size. It has everything to do with connectedness and concern with the group.

This need to connect is shown by what happens to us when we are placed in solitary confinement. After a certain period, we go mad. Or we will be driven to suicide. By contrast, Jane Goodall observed a chimp that lived apart from his troop for five years. He did not go insane. He did not die, and there was no reason to think he would. One of the most popular gorillas at Zoo Atlanta was without gorilla companionship for twenty-seven years. Yet he only became notably difficult during a period when his cage was being renovated, and he was placed in a crate so small that he could not fully stretch out his arms. An ape's identity is not so bound up with what others in the group think. In fact, gorillas and baboons are so lacking in social awareness that most can't even recognize themselves in a mirror. But manta rays, bottlenose dolphins, killer whales, Eurasian magpies, and some species of ants regularly pass a mirror test. In a number of critical ways humans appear to be unlike apes but akin to herd and pack animals. Like many dolphin species, we communicate through specific expressions, and we name our fellow creatures based on parentage.

This is not meant to disparage the work of primate field researchers like Goodall, Dian Fossey, or Christophe Boesch. They have provided us with an extraordinary amount of vital information about how primates in the wild behave, and academic theoreticians have offered us an array of intriguing notions about the sources and causes of human evolution. This book relies upon their research. This has been of immense value. But a critique and a corrective to their more extravagant claims is overdue. This is especially so as neuroscience and biochemistry are disproving so many of those assertions. In the end though we must look to behavior as the final test of how animals are alike or different.

The science of animal behavior is known as ethology. Arguably, Konrad Lorenz was the most influential of the twentieth-century ethologists. A native of Austria, Lorenz had a lifelong interest and fascination with animals. His most important and influential experiments were with geese.

Lorenz's research revealed something of much significance. The bird brain is designed to rely upon imprinting. This is distinct from learning. It is a bit like a predetermined computer code. Essentially, it tells an animal what to do at a certain time whether or not the creature has ever been previously instructed in the given task. Imprinting tells birds to make nests. Remarkably, there are species of birds that, though raised in a lab from when they are hatchlings, will make nests even though they have never seen one. (At any rate this is so if they are provided with straw and the other ingredients needed for the "recipe.") Lorenz also demonstrated that the instinct among geese to follow their mother is so great that when he deprived them of their mother he was able to convince them that he was she and then lead them about. Some psychologists took this latter discovery and tried to apply the concept to humans, but it's not exactly clear what the relevance might be given how different we are from mallards.

Other ethologists drew a different conclusion from Lorenz's research. They decided that an animal's capacity for problem-solving must be dependent upon the creature's class, order, and family. In this way, they came to believe that birds must be poor problem-solvers since their brains rely so greatly upon imprinting. The bird brain was thought to contrast unfavorably with the mammal brain as our class of species places emphasis upon learning. A further inference was that we could assume that as invertebrates don't have developed central nervous systems like ours they would have little or no capacity for solving intricate problems. This view of the animal kingdom provided a framework from which one conclusion unavoidably arose: Since humans are so exceptional at solving problems, then it must be that apes stand near the top of the pyramid, just below us. In a sense, this was an abstraction of that chart showing humans arising from apes.

This supposition was supported by the discovery that chimps and gorillas can make tools. Nearly all primate researchers report that at one time or another they have seen chimps and gorillas employing found objects in imaginative ways. Thus, de Waal wrote of seeing chimps using straws to get a beehive's honey, and Goodall had the opportunity to watch a chimp use a leaf as an improvised cup with which to hold rainwater. Taking all this together, anthropologists arrived at the idea that primates must be the smartest and most advanced of all the animals.

But primates are not necessarily smarter than other creatures. Nor must mammals be more clever than birds, or even invertebrates. It turns out that this line of reasoning is more than a little misleading and often entirely erroneous. Although octopuses have no cerebrum, they have large brains and are good at solving problems. For instance, if you leave them overnight in covered tanks in a darkened lab they will take the lids off their tanks, invade other tanks with fish in them, and eat them. Then they will return to their tanks in the early hours of the morning, affix the lids back on, and pretend that they had never left. Cuttlefish, a species of mollusk, have a brain that goes through their digestive tube. Yet, possessed of one of the largest brains among invertebrates, they are able to count. Similarly, crows excel at solving an array of difficult problems. Indeed avian researchers have shown that corvids can trick insects to get them to come out of hiding places, and they can quickly solve multistage puzzles to obtain sticks they will use to pry out food. And, as we all know, parrots can mimic human speech. In fact, one is reputed to have learned 950 words and to have figured out tense. Moreover, humans and other mammals display signs of instincts that are alike to imprinting. Fear of snakes seems to be unlearned. Many babies possess it innately. In the same way, rats become frightened when they scent cat hair, even when they have never encountered one before.

It seems that the mammal vs. bird vs. invertebrate model for understanding abilities at problem-solving doesn't work especially well. How easily an animal solves problems is dependent upon its skills at listening, cooperation, focus, motivation, inquisitiveness, brain size, and many other factors. That chimps and gorillas are genetically closer to humans than crows, cuttlefish, and octopuses tells us little about their capacities for problem-solving. And what we even mean when we discuss intelligence is open to question. That I may have the ability to use words in order to make an argument tells us nothing about whether I can sing or fix a car engine. (I possess neither skill.)

The first intelligence tests were developed by a Frenchman named Alfred Binet. Binet's tests required no ability to read or write, and he thought that they would be useful predictors of how well people could perform simple tasks. But, while they did a good job at indicating academic skill and suitability for certain elite occupations, they did not say how well people could drive a bus or clean a clogged pipe. Illiterate housemaids frequently turned out to be better at fishing than nuclear physicists. Brilliant pianists made poor farmworkers.

Chimps and gorillas have larger brains than dogs, though much smaller ones than people and dolphins. As we have already noted, there appears to be a relationship *among* primate species of problem-solving ability to brain size, but that does not even prove that this is true in a comparison *across* families and classes of species such as that between dogs and chimps. Though chimps have brains that are relatively and absolutely larger than those of dogs, it is not clear that this bears much relevance to the nature of human thought and behavior and our degree of kinship with great apes, and as we have already seen dogs are better at many tasks that human life requires than chimps are.

Jacques Mallet du Pan was one of the eighteenth-century French politicians who opposed Louis XVI's overthrow. It's to du Pan that we owe the expression, "Like Saturn, the Revolution eats its children." Although the French Revolution was bloody, du Pan was employing a metaphor, of course. Yet, horrible though it may be, many species engage in it. That includes filial cannibalism. This is when a starving mother consumes her own eggs. Not so uncommon among reptiles and birds, the practice reflects the low value that she places on the survival chances of any one of her eggs.

Filial cannibalism is less common among mammals, but it does occur. In particular, it frequently takes place among rodents. Less often is it observed among primates. For many years it was known that on rare occasions tamarins, bush babies, macaques, baboons, and orangutans eat their young. Then, within the last decade, primatologists witnessed chimp and bonobo mothers consuming their infants.

Chimp and bonobo mothers develop intense bonds with their children. So this may seem counterintuitive. Granted other facts though, it begins to make sense. Let's go back to the relative absence of meat in the chimp diet. Even cooked, ingestion of meat has damaging effects. Heavy intake of it reduces life expectancy. But it offers enormous benefits during the most productive years of life, those in which we are most likely to become parents and raise children. Meat has great numbers of calories. Perhaps even more importantly, it's the best source for complete protein: meals with every amino acid in them, presented in a balanced form. That helps in building and repairing damaged tissues and for overall growth. On top of all that, it's a superior source of B-12, a necessary vitamin for brain function and hemoglobin and DNA production. It is to be expected then that people who are not getting any animal protein in their diet—no dairy, fish, poultry, meat, or seafood—sometimes say that they feel starved.

This is the nearly permanent condition of chimps. They have each of the innate deficiencies as hunters that we do, and, as they aren't highly cooperative, they aren't especially skilled at it. At times, this must awaken desperation. What then do they seek out? Much of their supply of protein comes from the consumption of nuts, and it's surely no accident that one of the forms of tool-making that they have shown proficiency at is breaking up their casings. It's also a reason why they have huge jaws, since they use them in pressing down and shattering their shells. When they're really starved, though, they will target the prey least able to defend itself: babies. Given the chance, they will grab human infants. De Waal described one such incident:

> [It] involved a local woman, her baby, and her niece. The niece carried the woman's fourteen-month old baby. They were crossing a small gully when they ran into Frodo, who was feeding on oil palm fronds. When the ape turned around it was too late to run. Frodo simply plucked the baby off the girl's back and disappeared. Later he was found eating the baby, which by now was dead. Baby-snatching is an extension of predatory behavior and had until then been reported outside the park only. In nearby Uganda, it's become an epidemic, with human babies being snatched from homes.

To the same degree chimps like to target baboon infants. They will not, though, attempt to kill and eat adult baboons—no matter that they are anywhere from two to five times larger than baboons. But, lacking in our taboos, prohibitions, and law, in dire circumstances they will consume their own children: those least capable of fighting back.

This brings us to a larger truth. Evolutionary biologists refer to animals that have few offspring and that invest great amounts in parental care as K-selected. These contrast with animals that produce many offspring and invest less in rearing, the so-called R-selected. In general, animals that are K-selected animals mate more often, while R-selected animals may go through the process of reproduction only once in their entire lives. R-selected creatures have other common traits. On average, they tend to be smaller, live shorter lives, are subject to predation by other species, and live in unstable habitats.

Mice and rabbits are R-selected. This type of creature is more often encountered, though, outside the mammalian class. Oysters, bacteria, mosquitoes, and jellyfish are R-selected.

The distinction between K-selected and R-selected gives us a further indication that in vital respects we have stronger patterns of evolutionary connection to large-brained herd animals like horses, hippopotami, elephants, and cetaceans than primates. Both humans and primates tend to be thought of as K-selected. So, as well, are cetaceans. Yet there are ways in which human patterns of survival and reproduction match far better to those of whales and dolphins than they do to chimps. This requires some explanation.

The term R-selected comes from *reproduction rate*, while K-selected arose from *carrying capacity* (the number of offspring that an environment can contain). Studies of the birth rates of human hunter-gatherer populations do not suggest that women have evolved to bear fewer children than chimpanzee mothers. In fact, the lifetime fertility levels of women in hunter-gatherer communities are probably not meaningfully different from those of chimpanzee females. From this one might conclude that we are no more of a K-selected species than chimps and gorillas.

These terms offer other implications though. It is an axiom that R-selected animals have faster rates of growth and maturity. Thus, R-selected animals are referred to as *precocial* (rapidly maturing) and K-selected animals as *altricial* (slow-growing). This is how the difference between humans and apes manifests itself. That can be seen in our contrasting patterns of childbearing and child-raising. Chimp females typically have their first period and their first child four years earlier than women. Then chimp mothers continue bearing children into their forties, the time during which they reach peak fertility. Yet, on average, they will only have a child every fifth year. Women in hunter-gatherer societies carry to term more often, but there is a narrow window in which they are bearing most of their children. This is between a woman's twenty-fifth and thirty-fifth birthdays. After that, fertility rates—among women and the big-brained species of dolphins—fall precipitously. From that point on, their focus shifts from having children toward caring for them.

CHAPTER 8

DO YOU THINK HUMANS ARE SEXY? HERE'S WHY YOU DO

IF HUMANS AREN'T REALLY PRIMATES, then most of our thinking about human sexuality would have been distorted by the belief that we are. Is that so? A vast amount of evidence has come out in recent years to suggest that it is. Much of this new knowledge comes from genetic testing.

We'll look at that data. Yet this is not our only source of information. The research of the primate researchers provides us with other important clues. For a start, think about how we determine physical attraction. While it may not be what we wish, men think about a woman's sex appeal based in considerable measure upon her age. That is because female fertility declines over time. Yet primatologists have found that male chimps don't act or think this way. They are more interested in older females than younger ones, and anthropologists have noticed that female chimps in their forties have the highest fertility rates. To that degree, Jane Goodall reported that one of the female chimps she studied who was most sought after by males was so old that she was missing all of her teeth, and she had to eat her food with her gums. But she was highly fertile, and as Goodall observed she had enormous sex appeal. When she was in heat, there was a line of males waiting for her.

Humans are sufficiently malleable that we can certainly imagine this degree of attraction to an older female sex symbol like Madonna or Sharon Stone. However, in saying that we are again acknowledging that how humans think about sex is different from the way that chimps do. This is not only because human females don't go into

156

estrus. It's because much of our interest in sex symbols is in some measure a response to celebrity, a distinctly human concept.

Moreover, while women tend to synchronize their ovulation cycles with others in their group, chimps incline toward asynchronous estrus cycles. This is to say that they will naturally make themselves fertile when other females in their troop are not. It's believed that this is an evolutionary adaptation whose purpose is to avoid conflict among the chimp females. This is not exclusive to chimps. No primate synchronizes its cycles of fertility. That contrasts not only with our nature but with other highly intelligent herd animals. Elephants, for instance, are alike to us in that they tend toward synchronization of their periods of fertility, and dolphins, sheep, and goats can be artificially induced toward synchronization with hormone supplements.

But it's not only that human females are different from chimp females in our unconscious sexuality. How women judge sexual attractiveness is different from female chimps. Studies of online dating show that women place a high value on sense of humor. They also esteem income and wealth. Primatologists tend to equate income and wealth with the possession of status held by an ape alpha male, yet since chimps and gorillas don't have the ability to store goods or build homes, it's not clear that these attributes are terribly similar. One might add that how a man is perceived in terms of fierceness may have little connection to his income. A woman may be infatuated with a man, drawn to him for his money, but think that he is milquetoast. What is considered desirable in our time of tech billionaires is quite unlike what a female chimp finds sexy and alluring. Money conveys status. But it is rarely acquired through the swaggering displays of bellicosity and aggression by which chimps gain status.

In addition, such mannerisms are turnoffs to many, if not most, women, and, while women tend to prefer taller men, it turns out that men who are taller than six feet two inches are actually less desirable on average for women than men of less stature. That may be because human females seek connection far above the implied capacity for physical dominance that determines status among apes. Hence, we compose poems and sing songs to show our love, and there is little that we think of as more romantic—more intrinsic to human expression—than this display of affection, commitment, and desire. Similarly, young women are usually far more enamored with talented singers and musicians than the most powerful leaders. And while men may think that hugely muscled physiques are irresistibly enticing, one need only make a list of male sex symbols to see this is not so. From Paul Newman to Leonardo DiCaprio to Harry Styles, they have most often been slender, not hulking.

This principle seems to apply to some degree with respect to relations among men as well. One of the first pieces of advice routinely given by dress consultants to tall, strapping men is not to wear charcoal suits; they are being told to deemphasize their size and strength, lest they intimidate their coworkers. To the same degree, human sex acts are different from those of chimps and gorillas. As we have already mentioned, chimps and gorillas do not kiss during the act of lovemaking. The reason that we engage in that very human act we all know: pair-bonding.

There are a number of primate species that engage in pair-bonding. Some gibbons, titi monkeys, and owl monkeys form up into families composed of an adult male and female. In many cases, these prove to be lifelong bonds, and in these species, the males assist in child-rearing. The case of gibbons would appear to be particularly relevant as they are apes, albeit "lesser" ones. Often referred to as *Hylobates*, they are among the best of the primates at swinging from tree to tree. Their skill at this is so great that in a single motion they can fling themselves almost fifty feet while traveling at speeds approaching thirty-five miles per hour. This should tell us what their diet is composed of (leaves) and where they live (tropical forests). It almost goes without saying then that they are exceedingly territorial, and separate couples rarely cooperate with one another. In fact, they are legendary for the strength of their calls warning other *Hylobates* away. These shrieks and whoops are so loud that they can be heard as much as a mile off.

There are other mating patterns among primates that seem to have similarity to human ones. For example, some species of monkeys—like tamarins and marmosets—engage in polyandry. A dominant female breeds with a group of males that serve as her harem. Other females suffer acts of aggression and are prevented from attracting mates or bearing children. I think it's safe to say that this type of polyandry is different from that observed among humans.

But, in one respect, we do find something alike to human sexual practices among bonobos. *Pan paniscus* will sometimes engage in face-to-face sexual encounters. This is more akin to how we mate as the partners must be looking at one another during copulation, and there must be some intimacy during copulation.[4] This has never been observed, though, among *Pan troglodyte* chimps, and there is just one documented report of it among gorillas. Humans, of course, can have intercourse in either fashion. That brings us to something else that is relevant to understanding our sexuality: the design of our sex organs. Their evolution is reflective of ours, and they offer us additional hints that we are not rightly placed within the primate order.

Human females possess a hymen. It is commonly supposed that it is part of the vagina. In fact, it is a layer above the vagina that seems to function as a barrier to deter the invasion of harmful microbes—and male appendages—prior to sexual initiation. Hymens are found in many mammal species, including whales, dolphins, even-toed ungulates, and rats. However, chimps and lemurs are the only primate species possessed of them.

Similarly, humans are distinct from all primates in that we are lacking for an *os clitoridis*, or clitoris bone. The clitoris bone is the female analog of the penis bone. Often referred to as the baculum or *os penis*, the penis bone is present in many mammals. In fact, it can be found in all primates except lorises, spider monkeys, and woolly monkeys. Yet, though most primates have a penis bone, the list of the other mammal species devoid of one is long. No penis bone can

---

4   Cetaceans mate in the missionary position.

159

be found in any type of ungulate (hoofed mammal), elephants, marsupials, cetaceans, sirenians, lagomorphs (rabbits), hyenas, monotremes—and people.

It's been suggested that the penis bone permits the male of a species to remain erect and to have sex for longer than a creature absent of one. This argument depends upon the fact that while we may call an erection a boner, the human male equipment is made up of skin cells. Thus, the supposition is that this bone provides greater physical support for more sustained intercourse. Other theorists maintain that the penis bone permits its owner to place more sperm into the vagina or to settle it in deeper.

However it works, it seems to have an effect. A pair of Australian researchers demonstrated that through experiments with mice. When they placed male mice in a setting where they were not competing for females and they were permitted to reproduce freely, over a series of generations the penis bones evolved to be smaller. Yet, if the mice had to compete for females, it stayed the same size. There is also research that the lack of a penis bone is correlated with the extent to which a species practices monogamy. If this is so, it would obviously explain why humans are without one and chimps and gorillas possess one.

The function of the *os clitoridis* is less clear, as is the role of the hymen. Common sense would suggest that the existence of the hymen serves to protect women from early sexual encounters and molestation. A hint that this might be so is provided by the development of the labia in humans. In apes, the labia gradually disappear as the animal matures and passes through puberty. In humans, the labia get larger during puberty, eventually covering up and partially hiding the clitoris.

There are at least two other things that are peculiar about human sex organs. The first relates to the vaginal canal. In almost all mammals, it's horizontal. That is to say that it runs from the pelvis toward the tail of a four-legged animal. But while the canal starts in humans along the pelvis, it exits in front. The other mammals that have this peculiar design are the cetaceans, the pinnipeds, and, once again, elephants. This shift in the direction of the vaginal canal means that the baby must make a sort of right turn as it exits the uterus. The passage is longer and more difficult. As with the development of the labia and the hymen, this may be the evolutionary purpose as it provides added protection, placing the developing fetus farther away from microbial invaders. This is consistent with the idea that each human baby is rarer and of more value for evolutionary survival—a characteristic of a highly K-selected species. This might also explain why aquatic mammals and elephants have followed a process of convergent evolution of their vaginal canals. In the case of the cetaceans and the pinnipeds, this makes particular sense as the longer, hook-shaped birth channel serves to keep the child away from the swarming multitudes of microbes present in seawater. And, as we know, aquatic mammals and elephants tend toward longer lives, they mature slowly, and they bear relatively few offspring—like humans.

The final obvious difference between our sexual features and those of primates can be seen in a woman's silhouette. The region around the nipples of female primates will expand when they are lactating, but they do not have permanent breasts. In fact, no other species has permanent breasts. There are many theories to explain this. Desmond Morris argued that breasts were a way to attract male eyes as a substitute for female buttocks, which males once trailed when our ancestors were quadrupeds. For a number of reasons, I think we can dismiss this explanation. Possession of breasts involves a considerable evolutionary cost; large breasts can cause serious back problems, and they make women slower in running and walking. Neither is a minor matter given that humans have repeatedly neared extinction. What's more, we have been bipeds for four million years so men can't be effortfully seeking out a visual replacement for something they haven't set their eyes upon through two hundred thousand generations. It also seems to reverse matters. What's important isn't male awareness of feminine presence, but female competition for males. It's more likely that women possess breasts as a means of "costly signaling" of their health than that they are indicating that they are females. Any male can identify a woman simply by the fact that she doesn't have much facial hair, but, unlike chimp and gorilla females, women are competing for male spouses. The breasts then may be intended to show good health through a proven excess of fatty tissue.

That makes sense since adipose tissue makes up most of the weight of a woman's breasts, and it's a useful reserve of energy that can be called upon during strenuous times. Those extra calories to draw on in periods of famine can be vital given what we know of human evolution. That includes those repeated near extinctions and that we evolved as hunter-gatherers confronted with unpredictable food supplies. In other words, permanent breasts may have first appeared to sustain women when food was scarce, but they are now desired by men as a seeming indicator of overall fitness.

This would be an example of what is known as sexual selection. This is the evolutionary process in which males and females pick members of the opposite sex with which they wish to mate based on something that is merely presumed to be of survival advantage, or which is a sign of health or beauty, like the peacock's plumage. A related idea is that men see breasts as an indication that a woman can provide milk for their child, and men have therefore developed an evolutionary impulse to find females with bigger chests sexy.

This is what we referred to earlier: costly signaling. Secondary sexual characteristics are typical of animals that have to make an effort to stand out. That's probably why the primate species best known for its secondary sexual characteristic is the mandrill. They live in hordes of as many as 1,300—more than any other primate species. Within such an enormous agglomeration of individuals, the male mandrill must make an effort to be noticed by the females, and, perhaps for this reason, the males' faces are festively arrayed with bright red and blue stripes.

A final theory about why women have breasts is founded in research showing that women with more feminine figures—large bosoms, small waists, and full hips—have higher estrogen levels and are more fertile. That does not explain, though, why women continue to have breasts after menopause. That fact suggests that breasts do serve as a reservoir of extra fuel that may be called upon in time of crisis. This would explain why women and men have added stores of fat around our behinds. As dieters know, this is among the hardest weight to lose, and, unlike the muscles of the *gluteus maximus*, it is not all required for support of the torso and pelvis or for locomotion. Nor does the claim that women with pronounced figures are more fertile explain why this should be so for humans but not primates. Hence, we are led back to our original explanation: As humans are nearly the only animal classified as a primate without a steady food supply taken from leaves and ripe fruit found year-round on trees in tropical forests, we need more adipose tissue for survival.

That inference fits with a finding of researchers in Japan. Japanese snow monkeys are one of the few primate species that can be found in a cold region, one that lives where food is hard to obtain in the winter months. Consistent with this, Japanese researchers report that female snow monkeys are prone to obesity. This is a condition that is almost unheard of among primates. Yet investigators found that two-thirds of the female snow monkeys they examined met the clinical standard of being overweight.

This again shows how different humans are from nearly all primates. Because they evolved for life in the tropics and don't need to make dramatic adjustments in response to seasonal shifts in the availability of things to eat, primates don't store much energy, and they have extraordinarily low body fat ratios. For example, captive bonobos have average body fat ratios below 1 percent, and wild female baboons average body fat ratios below 2 percent. These levels would be taken as a sign of starvation in humans.

Those differences in appearance between humans and apes correspond to differences in how we sound and speak. The voices of male and female chimps are extremely similar. Insofar as there is any difference, it's the opposite of what you might expect: male chimps have slightly higher-pitched calls than females. Humans don't match the pattern of primates in that our male and female voices are so distinct. This also appears to be a form of costly signaling, and, like all costly signaling, it suggests something real: Men with lower voices have higher testosterone and cortisol levels. These indicate healthy immune function.

These aren't the only sex differences between chimps and humans. Another has to do with competitiveness within the sexes. Read a book on primate behavior and you will come to notice something curious: Mentions of jealousy tend to crop up regarding male sexual competition and not much else. Primatologists have searched for other examples of ape jealousy. To that end, de Waal performed studies in which chimps saw rivals being given nicer meals than they were. This provoked hostility and a sense of grievance. Commenting on this, he remarked, "If you ask me if there is a difference between the human sense of fairness and that of chimpanzees, I really don't know anymore."

Let's suppose for a moment that he's right. This hardly demonstrates that chimps have a similar measure of jealousy as people. After all, jealousy and envy aren't spurred merely by a concern with inequity. People experience envy because they have a poor sense of self, and they wish to be someone else or to be sitting in another person's seat. Absent of a strong sense of what their place or identity is, people fixate on someone else's possessions, qualities, or rank. This may be unrelated to conceptions of fairness. It has everything to do with social awareness and self-confidence. In its pathological forms, lack of self-confidence leads to conditions like anorexia and bulimia. It's probably no accident that these are uniquely human afflictions, and they are especially likely to strike adolescents: those who have little sense of accomplishment drawn of life experience.

A clue that chimp jealousy is different shows up in a passing comment de Waal made. Without reflecting on the significance of the observation, he remarks that jealousy in chimp females is "markedly" less noticeable than it is in male chimps. This makes sense as chimp females aren't rivals in the way human females are. No chimp female is competing with any other for a prospective spouse or a potential father who will assist in raising her children.

Now let's stop for a moment. Primate researchers have paid almost no attention to the idea that breasts might be a form of costly signaling, and they have been hard-pressed to explain the "markedly" less notable jealousy female chimpanzees display toward each other. In both cases, this is because it required them to think of human behavior and its emphasis upon monogamous mating patterns. They did not do so because they were starting out with the idea that we are not so different from chimps and gorillas. That was the problem: Those who try understand human nature from the behavior of apes miss at least as much as they grasp. That is also why they have ignored the fact that chimp female fertility cycles tend toward asynchrony while human and elephant fertility patterns are synchronized.

As an example of this problem, take the writer Wednesday Martin. In 2004, she married a private equity executive and moved to the well-heeled enclave of New York's Upper East Side. Her experiences prompted her to write a memoir in which she interpreted what she saw around her from the perspective of current notions in anthropology. The result was the book *Primates of Park Avenue*. It viewed the competition among "the ladies who lunch" showing off their Birkin bags in the manner of baboons seeking to establish status within a baboon harem. The problem with this is that baboon social structures and behavior don't correspond much to those of humans. Martin wasn't part of a harem. Nor was she compelled to have intercourse with multiple members of a posse of powerful, allied males who shared breeding rights over her. Nor did she live in a troop of fifty or more creatures who clambered up Central Park's trees, hopping from branch to branch searching for leaves. At no time was she faced by threats from leopards, lions, and crocodiles. So she never had to hide in the upper reaches of the forest canopy. In Martin's set, the abiding fear was that your husband might have a mistress, a prenup, and a skillful divorce lawyer. These were the dread events that undermined your social standing. Competing with one another for exclusive breeding rights with successful males hoarding plentiful resources, she and her friends were highly monogamous. And when their husbands were wiped out in market crashes and replaced in the executive suite by younger males, the newest men to rise up didn't grab their children from their lactating breasts and murder the newborns. Yet in some years baboon troops will see three-fourths of their infants rubbed out this way.

To the same degree, Martin probably didn't kidnap the baby of a lower-ranking frenemy in her neighborhood. That's a common behavior of female gelada baboons—so much so that a guidebook from the European Zoo and Aquaria Association warns zookeepers of the importance of making sure infants are with their actual mothers and haven't been stolen from them. The jealousy female gelada baboons display is mostly over having babies, not sole possession of rich husbands who are their fathers. The baboon model fails for other reasons as well. More than likely Martin probably spent some of her free hours teaching her children, but baboon mothers expend almost no time engaged in this task.

A better model for the behavior Martin observed might be a creature like a gray wolf or a coyote. These animals eat meat, they are terrestrial, they mate monogamously, and they jointly rear their young. The females compete with one another for the capable males. Female social relations can be complex as they team up with other she-wolves and female coyotes—often their sisters or other near relations—in raising their brood. The female coyotes set up a den while the males hunt, bringing back food for the pups. The mother rears the pups and teaches them the arts of hunting. Similarly, she-wolves take their pups to "rendezvous" sites where they can practice hunting as they approach maturity. Doesn't this sound more like the Upper East Side matrons who jointly socialize when they take their children to the park and wait for their husbands to return in the evening?

Another helpful model might be a highly social bird species that stocks away its food for the winter, like a jay or a rook. Rooks are able to make tools, and by a number of measures they are better at problem-solving than chimps. Male rooks spend the springtime courting the females. Then, once they are attached as a monogamous pair, the female rook will create a nest, and the couple work together to store resources for the brisk, sunless days ahead. Rooks also do as Martin's husband did: They hoard. However, in the avian universe, it is usually referred to as caching, and it is regularly witnessed as well among woodpeckers, nutcrackers, and jays. But, again, it is not a behavior encountered among chimps and gorillas. There are a number of reasons for this. Among the most obvious is that most creatures engaged in hoarding are sedentary, which chimps tend not to be. It is also characteristic of animals that live in seasonal latitudes.

All this takes us back to a basic point: Prior to the 1970s, few people thought much about human nature in terms of our classification as a primate, and prior to the 1990s no one referred to chimps, orangutans, and gorillas as hominids. Then, as these ideas became widely accepted, many patterns of human action became less intelligible. More than a few became inscrutable. By assuming that our distant primate ancestry means we must be like chimps, baboons, or gorillas, we aren't improving our understanding. We're diminishing it. If we try to understand humans as a "third chimpanzee" as author Jared Diamond would have us do, we are less able to grasp the motivations behind suicide and anorexia. But if we dispense with the idea that humans should be thought of as primates, then our view of human actions better aligns with evolutionary theory—and the evidence all around us.

On this basis alone we ought to reject the idea that our evolution corresponds terribly well with creatures that climb trees to get hold of leaves and berries. But that is not all that is wrong with this assumption—and it is merely an assumption, if a prevalent one—that humans are best conceived of in terms of our distant primate lineage. As we have seen, this notion is contradicted in many other ways. For not only have we evolved to be heavy meat-eaters but also speaking, co-breeding, domesticated animals. Recall then as well that while we are descended from primates; we are much more social, cooperative, and altruistic than any primate; and we have the sleep patterns and the qualities of endurance and perseverance of such herd animals as horses and elephants. Only by acknowledging all that do things start to make a measure of sense. The primate habitat and diet are not like those of humans. They are very much akin to those of sloths. In the rainforests of Central America, they occupy the same trees, and their diet tends to be very much alike.

Animals adapt to their habitat and their diet. Primates and sloths are very different. Sloths are in between cold- and warm-blooded animals. Their normal body temperature is just eighty-five degrees Fahrenheit. They rely upon the sun for body heat, and, as we all know, they are famously slow-moving. Like gibbons, they are solitary and territorial. Three-toed sloths share the vegetarian diet of gorillas. Two-toed sloths are like chimpanzees in that they will supplement their consumption of leaves with amounts of fruit, insects, and even, on occasion, small vertebrates. Both types of sloths are alike to apes and monkeys in that they can serve as a meal for big cats. This is why they hide out in the upper branches of trees, where they spend much of the day sleeping. Yet they appear to have a set of evolutionary needs and a resultant nature that place them much closer to chimps than chimps are to humans.

Trying to understand people through the study of apes, evolutionary biologists have not only misunderstood the functions of our body parts but also the act of sex. That's because ape sexuality is different from human sexuality. They are about as much alike as a banana is to a wind turbine, or a perfume bottle is to a steam shovel. For chimps, sex is almost exclusively a procreative act. But study of humans suggests that its evolutionary function for us revolves around its use in strengthening pair-bonds. And plainly that is of vast importance in terms of getting males to participate in the rearing of the young and thus in human survival.

A little more than a fourth of primates are classified as "socially monogamous" animals. This term does not mean that the creature is faithful but rather that the species arranges its living patterns alongside another of its kind. As psychologist Christopher Ryan put it, the expression is not meant for "inferring any sexual interactions or reproductive patterns."

Ryan is the coauthor of the popular bestseller *Sex at Dawn*. Its subtitle is *How We Mate, Why We Stray, and What It Means for Modern Relationships*.[5] Its authors wrote it under the influence of the idea that human behavior is very much like ape behavior, and the book typifies the sorts of errors that will arise when we employ this line of thinking. Its message is that it is possible for humans to be routinely unfaithful, but that this will not have unwelcome consequences for child-rearing or society in general. To that end, it approvingly quotes a primitive tribesman in northeastern Quebec who told a seventeenth-century French missionary sent to convert him to Christianity, "You French people love only your own children, but we love all the children of our tribe."

In support of their claim that sexual monogamy is neither natural nor desirable for our species, Ryan and his coauthor, Cacilda Jethá, provide the reader with assorted snippets of information about apes. Alongside this are examples of the activities of a number of polygamous tribes from around the globe. This is variously stitched together in order to present their case. What Ryan and Jethá fail to acknowledge is the vast amount of data on what polygamy does to the quality of child-rearing. That is because it does not suggest that polygamists "love" all their children—or at least that they do not raise them with an abiding record of success. We are told that there are remote polygamous tribes to be found in select pockets of the earth, but not what their conditions of life in these places are. Thus, Ryan and Jethá neglect to inform us that the Canadian tribe they touted was deeply troubled and that it has remained so. Ordinary public health statistics

---

5   The book was originally subtitled *The Prehistoric Origins of Modern Sexuality*. Possibly because psychologists and anthropologists were almost universally scornful of it, it was given a new subtitle and subsequently marketed as a "personal growth" book, and it is now generally to be found in bookstore self-help sections.

bear this out. One health researcher who studied natives from the region in the 1970s found infant mortality rates were 60 per 1,000. One-fourth of all deaths were among children.

This is a worldwide pattern. Polygamy is consistently associated with familial strife and haphazard parenting. In Sierra Leone, infant mortality is 64 percent higher in polygamous households than in monogamous ones, and in Mali researchers found that a child's likelihood of not living to adulthood is seven to eleven times greater in a polygamous family.

This phenomenon is not limited to non-Western societies. Indeed, a special proof of this may be seen in Mormon history. I think I speak for many non-Mormons in saying that I have met no group of people who form such a uniform impression of politeness, conscientiousness, or industry. Yet increased rates of early death in polygamous households were regularly observed by nineteenth-century visitors to Utah who studied their life and customs. Further, the small towns composed of members of the off-shoot sect of Mormonism that continues to practice polygamy are characterized by friction and distress. In order to sustain the system, the faithful are kept from going to college or even finishing high school. Boys are forced to leave the spots that they have grown up in when they become rivals to their fathers and uncles, and girls as young as twelve are pressured into marrying older relatives. In order to create the necessary degree of social conformity, they live in a condition of constant tension and forced obedience, and the leaders of the sect are more apt to be known for their criminal convictions than their corporate start-ups. Nearly all the members of the community are desperately poor.

The pattern of life that emerges is epitomized by that of the long-time head of the "Fundamentalist" Mormon Church, Warren Jeffs. When rivals within the church opposed him, he responded by "re-assigning" their wives to men who were loyal to him. On one occasion, he even "re-assigned" the wife of the mayor of the small town in which he lived to one of his allies. Jeffs eventually wound up on the FBI's Ten Most Wanted List. Testimony during his trial revealed that he had been raping a nephew from the time that the boy was seven and a niece from the time she was five. He is now serving a twenty-year-to-life sentence for child molestation. He regularly tossed out his own sons, forcing them to leave home when they turned twelve or thirteen, and he kept them from receiving an education, watching television, reading books from outside the home, or even owning toys.

This is not such an extreme example. All polygamous societies exist in a condition of either extreme mistrust and friction—in which the males are in conflict or carefree destitution in which they are persistently uninvolved in parenting. There is no advanced society anywhere in the world based upon anything except monogamy.

This is not all that is wrong with Ryan and Jethá's book. Using the example of bonobos, they argue that we have no reason to experience jealousy if our mate is sleeping with someone else. But we are not bonobos. What this is apt to lead to for our species is divorce or separation, and the replacement of a biological parent for a non-biological one. This tends not to be the equal of the tutelage provided by biological parents. One famous and much-cited study revealed that stepfathers were eight times more likely to molest the girls they were raising. The researchers found that one in six women brought up by a stepfather was molested. Yet the attitudes and beliefs displayed by Ryan and Jethá are characteristic of many of those influenced by a primate model of human sexuality. The book, in fact, is especially popular among people who self-identify as polyamorous.

Underlying their theories is the belief that unfaithful mating advances the evolutionary interests of both sexes. Males, it is suggested, who cuckold other men should be increasing the chances of the survival of their own genes. Likewise, it is assumed that unfaithful women will be eager to cheat with men with more desirable attributes than those possessed by their regular lovers; they will want to betray them with bigger, stronger, smarter, generally fitter specimens in order that their children will have these welcome traits. Operating with these premises, evolutionary biologists presume that such infidelity is common among humans.

The difficulty with this line of thinking is that it does not correspond with human behavior. Recent research reveals that our species has exceptionally low rates of "extra pair paternity." Women betray their mates and bear offspring from other men much less often than do females of nearly every other "socially monogamous" species. One team of researchers in Belgium, for instance, who examined a range of studies on human lineage found that the rate of female infidelity leading to conception by a man other than the woman's spouse rarely went above 3 percent. They observed:

> Perhaps, the most important finding of recent reviews of predominantly Western populations is that cuckoldry is normally very infrequent—in the range of 1.9% to 3.1%. Five subsequent studies on dissimilar populations also gave very low estimates of 0.94% in Germany, 0.73% for Afrikaners, 1.3–2.9% for different Dogon religions [in West Africa], 0.9% in Belgium and 1.2% in Italy.

Another researcher who looked at sixty-seven separate studies on the subject concluded that the average rate of false paternity in humans was just 1.7 percent. By contrast, in socially monogamous bird species the rates are often around one in five.

Because research with small samples can be unreliable, many researchers throw out studies with data taken from fewer than forty offspring. One paper that examined socially monogamous mammal species and their rates of extra-pair fertilization included thirteen species that met that standard. The median rate of false paternity for the group was 12 percent, and the average was 22 percent. These rates correspond to evolutionary theory. The problem as the idea has been applied to humans is the *infrequency* with which women conceive children with men other than those supposed to be the fathers.

Some women will repeatedly cheat on their spouses and lovers in an act known as extra-pair copulation but will deliberately not conceive a child this way. Either they fear the possible consequences of bearing a child from a furtive affair, or they are motivated by shame and a sense of decency. Whatever the motives, this is the opposite pattern of action from the one evolutionary biologists claimed women ought to be practicing, and within the last decade the real data has prompted researchers to suggest that the notion that false paternity is common should be, as a *New York Times* science writer put it, simply regarded as a "myth."

When we have regular sex with our mate, we develop feelings of attachment toward them, and this plays a central role in our habit of honest procreation. It's very likely that this is a consequence of our much greater capacity for empathy—a trait that we know tends to be wanting in apes. Plainly, though, there are other explanations to consider. The discovery of cheating in monogamous social animals can lead to ostracism. Since humans have some of the characteristics of herd animals and herd animals will do almost anything to avoid becoming outcasts, this is suggestive. One interesting finding with birds is that rates of extra-pair fertilization are highest among species with short life expectancy. This implies that humans may be more constant because we are more long-lived. Then there is the need for male participation in child-rearing and the desire of women to have confidence from their mates that they are the father of a child. Are all these factors in the human tendency toward faithful procreation? Are there still more?

These questions should be a primary area of concern for evolutionary biologists. But, working from the relatively indiscriminate mating patterns of chimps and gorillas and their lack of any instinct toward mutual child rearing, until recently evolutionary biologists were sidetracked. They focused on the possible causes of straying rather than the reasons for our constancy.

Ryan and Jethá's book *Sex at Dawn* may have been the apogee of this phenomenon in that it argued that humans are socially monogamous but not sexually monogamous. However, the opposite, it appears, is true. While we can be sexually promiscuous, this leads to the breakdown or even the absence of patterns of "social monogamy." Pair-bonding, good parenting, and a healthy social order result from sexual monogamy, and social monogamy follows from it, not the reverse. This is surely part of the reason why there has been a traditional hostility toward the legalization of prostitution. It has been assumed that because it offers an alternative enticement to sex with a spouse that it weakens the impulse toward pair-bonding. It is hard to imagine that this fear is awakened by the belief that men will regularly leave their wives for prostitutes but rather that there is worry about the effect on pair-bonding caused by extramarital relations.

In many cases sexual monogamy may be a stronger force in our social patterns than social monogamy is. Proof of this comes from a recent study that looked exclusively at poor urban slum-dwellers. It yielded a false paternity rate of 5.9 percent. The study's authors noted that these results were an outlier in the larger body of their research, which indicated overall false paternity rates nearer to 1 percent. The downtrodden study population they examined was one in which historical rates of out-of-wedlock births were high, single parenthood was common, and most of the putative fathers had not stuck around with their lovers while the children were young. In other words, the researchers were examining people who did not easily fit the categorization of being "socially monogamous." Even so, they still had accurate notions about their children's paternity in over 94 percent of cases. In other words, they appear to have been more sexually monogamous than they were socially monogamous. Once more, working from a primate model of human behavior, evolutionary theorists had gotten matters backwards.

One of the most powerful demonstrations that sex plays a central role in human pair-bonding has only appeared since gay marriage became legal. Anthropologists and sex researchers had feared that this might produce high divorce rates among gay men. They took for granted that lesbian couples would have low rates of divorce. Yet the elevated levels have turned up among lesbian couples, not gay male couples. Nor is this the case only in America. It's cross-cultural.

Thus, between 2005 and 2009, lesbian divorce rates in Belgium were twice those of gay men, and data likewise show double the divorce rate among gay women in Sweden and the Netherlands, while Great Britain's *PinkNews* reports that in 2019 divorce was three times as common among lesbians as among gay men in the United Kingdom.

You might ask: What's going on here? Doesn't this conflict with everything we have heard about men and women—and especially gay men and women? After all, men are furnished (and often burdened) by nature with a more relentless sex drive. They are inclined, as Nora Ephron put it, to have sex with a venetian blind when nothing else is available. So it might stand to reason that married gay men, absent the civilizing influence of women, would be the ones who would continually split up while lesbians would stay faithful. The respective attitudes toward domesticity versus wild abandon are reflected in an old joke: "Q: What's a second date for lesbians? A: Picking out drapes. Q: What's a second date for gay men? A: What's a second date?"

In fact, more plausible than promiscuity as an explanation for the frequent failures of lesbian marriages is a phenomenon referred to as lesbian bed death. Nearly half of lesbian couples have sex once a month or less. By contrast, 85 percent of married heterosexual couples have sex at least once a week. One might respond here by saying that the lesbian couples are very likely faithful, so aren't they sexually monogamous? But this is erroneous. All too often these couples have become *asexual*. Thus, once more, we see that for humans sex is a vital agent for pair-bonding, and the capacity of humans for social monogamy is driven by habits of sexual monogamy.

Alternately, one might question this idea by pointing to the recent publicity given to the purported phenomenon of "polyamory." More than a few young people claim that they are not naturally monogamous and do not wish to be. Granted its sensationalistic nature, this is something that the press has taken a great interest in, and if you live in a big city you may have encountered this, meeting a couple who say they are in an open relationship. What you actually find, however, when you observe those who claim to be polyamorists over time is that one of the pair—sometimes both—has become desperate and fearful. They are clinging to their mate, trying to hold on to them. For, while they might like the idea of an open relationship in the abstract, in their own particular case, they are jealous, watchful, and possessive.

Ryan and Jethá think that humans can be sexually indiscriminate in the manner of bonobos and that this will be as easy and comfortable as it is for apes. Perhaps, they suggest, it will be more healthy for people. We will be relatively free of guilt and painful feelings, and our lives will be void of larger complications. There is little evidence of this though. That may be why Ryan wrote in the fashion that he did to *The New York Times*, criticizing an article it printed that attacked his belief that monogamy is contrary to our nature. The article noted that there is anatomical evidence suggesting that humans may have developed monogamous mating patterns as early as 3.5 million years ago—even before the invention of fire. It also made the point that children are more likely to survive when their parents are monogamous. Rather than object to the facts it presented, Ryan questioned its terminology. This is far more revealing of dishonesty than any paternity test.

The fact that humans pair-bond would seem to be an advantage for women as they have men to assist them in the rearing of children. But it may be as much a threat to a woman's evolutionary advancement and her happiness as it is a benefit. Feminism has taught a generation of women that the middle-class Victorian insistence upon maintaining virginity until marriage reflected a male desire for control over women. This may be. But it also offered protection against the baleful effects of pair-bonding. After all, if people are naturally inclined to develop intense feelings for their lovers and this particularly affects women, then pair-bonding can waylay women, drawing them away from their best interests. We see this in our own time when women become involved with men who have no intention of marrying them and find themselves sacrificing their most fertile years in relationships that do not lead to a stable family life or children. In the same way, pair-bonding is problematic for men. If they are unfaithful to their spouses, it magnifies the risk that their infidelity will break up their marriage. And, just as it does to women, it can lead them into unproductive or unhappy long-term relationships.

There are other unwelcome consequences of pair-bonding. Male chimps and gorillas will try to block access to females who are in estrus, locking them into the role of their baby mamas. Yet when they are discarded by their mates, male chimps are not apt to become obsessive ex-lovers, stalking them while angrily complaining that their heartfelt messages were ignored. Because we pair-bond, we invest more into our love affairs. We rightly call them *love* affairs. This is a two-edged sword. As we all know, men sometimes kill the very women they say they love. A study in Louisiana found that more than one-fifth of all murders were the result of "domestic trouble."

At the same time, because pair-bonding appears to be one of our strongest inborn impulses and sexual monogamy is designed to facilitate this, prostitution cannot be seen in the manner that most Marxists regard it: as an ordinary commercial endeavor that's not so different from sex within marriage. In fact, it is an activity that tears prostitutes away from our species' needs for connection and attachment. More than likely that's why prostitutes have such high rates of depression and anxiety. Were humans sexually indiscriminate in the way that chimps and bonobos are, this would not be so.

CHAPTER 9

WHY PEOPLE COMMIT MURDERS, START WARS, AND BUILD PYRAMIDS

IF YOU HAVE READ THIS far, you have almost certainly come to grasp one essential message of this book: We are herd animals. That proposition leads us to the not so small question of why people kill each other.

Among the most common arguments primate researchers advance to persuade themselves—and others—that humans are like chimps and gorillas is that we murder each other. Yet, like so many claims made by the anthropological establishment, the claim proves not so impressive upon examination. As the author of a recent popular science magazine article noted:

> In 2016, researchers in Spain analyzed data from more than 4 million deaths across 1,024 mammal species, including humans.

> Given that there were nearly 16,000 human murders in the United States alone in 2015, according to FBI data, and a plethora of motivations people have for committing murder—from jealousy, to squabbles about money, to hatred toward those who are different—it'd be easy to think that Homo sapiens sapiens would be the species most likely to kill its own kind.

> But humans didn't even rank in the top 30, though other animals commonly thought to kill each other—

wolves, lions and nonhuman primates, including various monkeys and lemurs—did.

The research also revealed that a number of seemingly peaceful species are surprisingly murderous. Long-tailed chinchillas, ground squirrels and several ungulate species—including wild horses, gazelle and deer—all ranked in the top 50.

The most murderous mammal species? Meerkats—around 20 percent of meerkats meet their end at the hands (and teeth) of other meerkats.

THE MAMMALS MOST LIKELY TO KILL THEIR OWN KIND

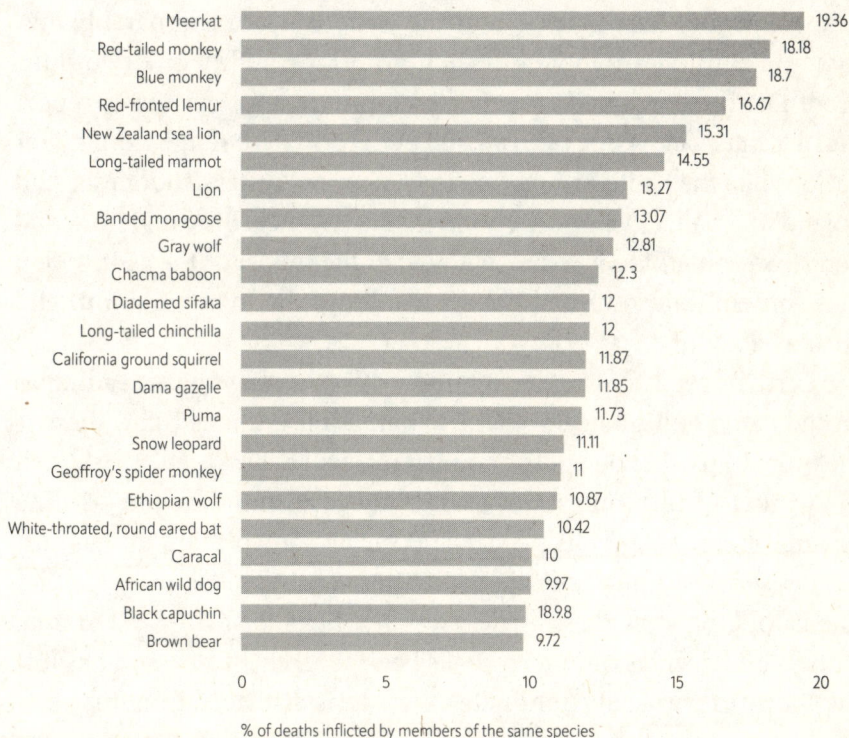

Species	%
Meerkat	19.36
Red-tailed monkey	18.18
Blue monkey	18.7
Red-fronted lemur	16.67
New Zealand sea lion	15.31
Long-tailed marmot	14.55
Lion	13.27
Banded mongoose	13.07
Gray wolf	12.81
Chacma baboon	12.3
Diademed sifaka	12
Long-tailed chinchilla	12
California ground squirrel	11.87
Dama gazelle	11.85
Puma	11.73
Snow leopard	11.11
Geoffroy's spider monkey	11
Ethiopian wolf	10.87
White-throated, round eared bat	10.42
Caracal	10
African wild dog	9.97
Black capuchin	18.98
Brown bear	9.72

% of deaths inflicted by members of the same species

Adapted from Nature.com No. 19758.

Yes, the researchers did observe that primates have a high murder rate, and we know that chimps are particularly homicidal. In addition, the researchers specifically noted that murder rates tended to be alike in related species. So shouldn't this be proof of how murderous we are?

For several reasons, this explanation doesn't work. First off, using our genetic ties to chimps and gorillas and the assumption that we fit with a primate model of behavior, the researchers calculated that 2 percent of humans should be dying by homicide. Yet few present-day human populations come anywhere near that rate. Instead, in some human societies murder is exceedingly rare. For instance, Japan had just 334 among its entire population of 125 million in 2021. That's about 1 percent of the number predicted. Nor is this something only seen among East Asians. Homicide rates in many rural communities in Europe, Asia, Africa, and the Americas are comparably low. In some countries people kill each other at roughly one two hundredth of the predicted rate. Forty countries and territories report murder rates below one per 100,000 per year. These range from major nations like the United Kingdom, Italy, Greece, Austria, Indonesia, and Poland to small, developing countries like Guinea-Bissau, Benin, and Tonga. So something is wrong with either the model or the assumption that contemporary human behavior follows the logic of chimp and gorilla behavior.

Further, even that one two hundredth rate may be misleading as an indicator of human impulses. When adult animals fight, there is the probability that one of the combatants will be maimed. Most often the winner of the contest does not bother finishing the job—killing his foe—because there is no need. Typically, an animal that has suffered a serious injury will die soon afterwards. This is true whether one is looking at snakes, turtles, reptiles, bugs, or primates. Among chimps and gorillas once a rival has been severely beaten and has lost the ability to physically dominate others he is displaced from his position and deprived of his authority. There is rarely cause to finish the job. That chimps will go ahead and slay another chimp tells us about their instincts. This is consistent with what we know of their nature. What we might call the effective murder rate is far higher among many mammals—perhaps most—than it is among humans.

This gets us to a larger point. In her writings, Goodall marvels in describing an incident in which a bigger chimp permitted a smaller chimp to eat some meat he was strong enough to take away. That power among chimps is most often a simple arithmetic of brawn isn't only Goodall's view. In de Waal's book *Chimpanzee Politics*, he unabashedly acknowledged this. De Waal then qualified this judgment with his observation that it also depends upon alliances within the troop and some amount of bluffing in which chimps engage in gestures and movements meant to exaggerate their size and strength. But, in the absence of physical strength, there can be no authority, and no chimp can be an alpha male—the leader—who is small or puny. Studies of chimps in the wild confirm this, demonstrating that leadership position is primarily based on physical size, strength, and level of testosterone. There is no Napoleon complex among chimps as there are no Napoleons. There is no equivalent to tech geeks who gain supremacy though their manners are meek and their bodies are frail. Small chimps don't rule. Diminutive chimps don't win followers or do much mischief. Nor do any chimps or gorillas own other chimps or gorillas. Incapable as they are of our meritocratic forms of society, they are also unable to construct a state in which chimps and gorillas are purchased or sold or reduced to the status of serf, untouchable, or slave.

We are different. Consider our most powerful leaders. The four mightiest during the Second World War were Hitler, Stalin, Roosevelt, and Churchill. Roosevelt could barely walk, and by the end of the war Hitler and Stalin had crippled arms. And what of Churchill? He was a short, obese drunkard.

It has always been so. Tamerlane's name literally means Timur the Lame, as he was. Some of the greatest leaders have been blind in one eye. A list of such formidable commanders includes Enrico Dandolo, the legendary doge of Venice; Horatio Nelson, Britain's greatest admiral; Moshe Dayan, Israel's most admired general; and Mullah Omar, the Taliban ruler. Others, like Senator Bob Kerrey, were missing legs. Kerrey's fellow senator and presidential candidate Bob Dole also lacked use of an arm, while George Wallace was able to run for president even though he could not rise out of his wheelchair. Human power does not depend upon physical strength. Consequently, maiming or mauling an opponent is not a sufficient means to victory in a fight. Often humans commit murders because it is the only way by which to resolve a conflict or settle a score. Thus, people commit murders as a means of self-preservation. Having committed a lesser crime, we only feel safe after performing the greater one. Once we take this into account, we can see that humans are not an especially violent creature and many of our murders are unlike those perpetrated by chimps.

I mentioned in the first chapter of this book that there's a group of primatologists who are iconoclasts. Rebels within their profession, they have (nonviolently) attacked the conventional claim that humans and chimps are greatly alike. I would not place Richard Wrangham among them. But he can and does like to challenge conventional opinion. Hence, in *The Goodness Paradox*, Wrangham acknowledged the obvious fact that humans and chimps have been moving on very different evolutionary paths. Then he directs our attention to a number of the resultant disparities in behavior. As part of this, he sees the human practice of war as an example of what he calls "proactive violence." He distinguishes this from the "reactive violence" of chimps: sudden outbursts of aggression. Wrangham explains the distinction by giving the example of a crowded commercial airplane. He says that if chimps were packed onto such a flight there would be constant anarchic violence. For humans the danger comes from the sort of deliberate violence that terrorists engage in, and this compels us to arrange for elaborate security procedures when we are boarding planes.

Reactive violence explains the high murder rates in big American cities. Studies of these killings suggest that this is the single biggest cause of homicides in dangerous urban enclaves. Wrangham believes that humans are evolving away from this type of violence. Although I agree, I think that Wrangham is failing to see the distinctly social and human component in these acts. These fights are over reputation, and this is something that humans have engaged in throughout history, whether through vendettas, duels, or honor killings. They are disputes that do not arise over power or authority so much as around questions of status and loss of face. They are further proofs of the human obsession with what *other people think*. They are a dark underside to our sociability. Take as an example the case of Howard "Pappy" Mason. He was the mob boss who controlled cocaine and heroin sales in Queens, New York, at the height of the crack epidemic of the late 1980s. That yielded him millions of dollars each year in profits. Mason lost that because of his obsession with "rep." The trouble started with a seemingly trivial incident. Hanging out with some friends drinking a beer in a public park, Mason was approached by a local policeman walking his beat. Told to get a paper bag to put around his beer bottle so that he wasn't violating the city's open-container law, Mason asked the officer, "Do you know who I am?" When the cop wryly responded, "Yeah, you're the guy who's going to get a paper bag," Mason became incensed. In response, he decided to hire hitmen to murder the officer as he had been "dissed." The killing drew the attention of President George H. W. Bush, the press, and the public. With this came the unraveling of Mason's empire. Serving a life sentence at Allenwood Federal Penitentiary in Pennsylvania, Mason will likely die in prison. This is a kind of violence that is distinctively human. Carefully arranged, it was motivated not by the desire for position or advantage but because of a preoccupation with prestige.

In some sense, the 9/11 terrorist acts are alike to this. None of the killers gained anything of immediate value. Rather, they were ending their lives. Osama Bin Laden's preoccupation with the presence of American troops in Saudi Arabia and his belief that this required retaliation fall into the same family of acts inasmuch as his schemes were intricately planned but not spurred by obvious self-interest.

The largest and most thorough study that looked at why murderers commit their crimes was undertaken by American sociologist Marvin Wolfgang. After interviewing 600 killers in Philadelphia and offering them twelve different categories of possible motive, Wolfgang reported that the single most common cause they gave for their actions, representing 37 percent of all murders, was "altercation of relatively trivial origin: insult, curse, jostling, etc." They thought they had been "dissed."

What Wrangham is right about is that murderous human impulses can be awakened collectively. Most often, these are driven by irrational if commonly accepted notions. We have mentioned the widespread practice among ancient peoples of human sacrifice. The Phoenicians serve as an example. Sometimes called the "Purple People" because of their habit of wearing attire dyed this color, they were a remarkable nation. Creators of the first alphabet, they learned to travel on the open seas making use of the stars, and it's said their mariners circumnavigated Africa. Arriving at their decisions through representative assemblies, they set up colonies throughout the Mediterranean and fashioned a commercial empire extending from their homeland in modern-day Lebanon out to Spain. Other of their outposts appeared in Cyprus, Sicily, and most famously at the city of Carthage in Tunisia. Yet, for all these achievements, they were terrified of the cruel gods they worshipped, and they were anxious to propitiate them. Persuaded that child murder was the sole means by which they could avoid the retribution of Baal and Ishtar and fearful of what their neighbors might think if they did not go along with these rites, the Phoenicians regularly slit their babies' throats. Following a procession led by a priest, they would take the dead infants from an altar whence the children were placed on a pyre for immolation. The Israelites' awareness of this, and their abhorrence for it, is the inspiration for the Biblical expression "burnt offerings." Similarly appalled, Greek and Roman writers made repeated reference to the Carthaginian propensity for human sacrifice. Nonetheless, later historians, incapable of imagining such behavior, frequently dismissed this as a libel. However, study of their infants' preserved remains confirms the truth. Nor is this legacy unique. Forensic pathologists are finding evidence of bloody rituals of human sacrifice all over the world.

The ancient Greeks and Israelites were among the first to reject human sacrifice. In its place they offered up goats, lambs, and other creatures they reared. This represented a conscious substitution for the sacrifice of humans: older rites they had spurned. We know this from the story of the Sacrifice of Isaac recorded in the Old Testament, as well from the writings of Aristotle's most famous and accomplished student, Theophrastus. His *On Piety* makes the connection explicitly. For Greeks, the custom was to reserve the animal bones and the fat for the gods. The ancient Israelites showed their courtesy by performing sacrifices upon an altar in their high temple in Jerusalem: Solomon's temple. Set on the location of the present Wailing Wall, it is the most sacred site in Judaism. The list of other peoples who performed ritual sacrifices of animals is long. It includes the Scythians, Romans, some Hindu sects, all Muslims, and all followers of Santeria. In Pakistan alone ten million animals are sacrificed each year during the annual festival of Eid.

Archeologists believe that animal sacrifice may have taken place in the ancient Near East as early as 5,000 years ago, and there is evidence of it in China from the Shang Dynasty. That was more than 3,000 years back. Confucius, who lived in the sixth century BC, considered it sufficiently important that he made a long list of the procedures to be followed when sacrificing animals in his *Book of Rites*. These acts are to be performed with solemnity as it is taken to be part of the respect due to the divine.

While these rites reflected a deliberate rejection of the earlier practice of "burnt offerings"—human sacrifices—they still must seem a bit curious. After all, none of these societies were characterized by plenty. Israel and Greece are fairly dry, and the first Muslims were desert folk of the Arabian Peninsula. The sacrifices they made were *sacrifices*. They were yielding up food they needed. These are powerful proofs of our social awareness, our impulses toward collective action and obedience, the extent of our underlying fears, and of our sense of shame.

By contrast, though chimp killing is most often driven by rage, it is Machiavellian. Characteristically, it is undertaken with a specific egocentric intention. Generally, it combines frightening anger with cold-blooded calculation. The accounts of lower-ranking chimps overthrowing the alpha male in their troop and then eating him are examples. In these cases, murderous chimps are excited and inflamed. Their hair is standing on end as they attack, and they shriek and holler. But, amidst their rage, they are acting rationally, jockeying for higher position and greater status. This contrasts with the tale Jane Goodall tells of a group of young chimps who regularly kidnapped their mother's youngest offspring in order to get their aged mother to give up possession of a termite heap they desired. Here the action was utterly ruthless and remarkably ungrateful, yet it was stripped of emotion. In both cases though what we see is the ape's brutal rationality. Chimp males target other males or infants. They do this as a practical act in order to advance their opportunities to mate, to expand their territory, to boost their status, or to acquire food. Simply put, Wrangham is on point in his claim that humans are more dangerous intra-species than almost any other animal *because* we are tame and obedient. The question is whether he has fully appreciated all the nuances and implications of the distinction between reactive and proactive violence and what this says about humans.

In either case, while our capacity to dutifully engage in acts of mayhem may no longer be reflected in human sacrifices, it is displayed in other, grander ways. Thus, when we talk about the murder rate, we exclude the killings committed in war and by governments upon their own people.

That toll dwarfs anything done by the Aztecs or Carthaginians. Consider that seventy-five million deaths resulted from World War II. This includes twenty million battlefield casualties. Then there are the twenty million additional Soviet fatalities that Stalin had a hand in. That's a portion of the estimated one hundred million lives communist regimes ended during the twentieth century. Going further back, it has been suggested that Tamerlane was responsible for seventeen million murders. That was approximately 5 percent of all the people then on the earth.

Yet each of the leaders behind these bloodbaths believed that he was doing good. Stalin, Mao, and Pol Pot were trying to follow Marx's plan by creating true and faithful dictatorships of the proletariat. Hitler was seeking to aid the *Volk* in its quest for *Lebensraum*. Tamerlane believed that he was obeying the will of Allah. Alongside them, carrying out their orders, were depraved henchmen. None of this would have been possible if thousands of people hadn't been docile servants of their will and sincere and devoted followers of their faiths.

What we ordinarily think of as the rational cause for a murder—a fight between two individuals over land, power, or the chance to mate—is far more often fatal between chimps than it would be between humans. But, though chimps do engage in frequent small, local battles among their troops, they are never more than that. Absent of languages, they have no identity based upon it. Possessed of far less social awareness and only those ideas that can be expressed without words, they have little notion of injured honor born of minor slights and no ideologies or belief systems. While primate researchers continually refer to these conflicts as "wars," we must put that word in quotation marks.

In short, the great danger that our species presents to itself is not that we are like chimps but that we are unlike them. *We* become preoccupied with snubs and insults that have offended our sense of our importance. That can make us dangerous as individuals when we live in a lawless state or among a fraternity of lawless men. That is our capacity for reactive violence. But our impulses toward proactive violence are more concerning still. At one time that regularly prompted us to follow religious dictates calling for us to slit our own children's throats. If that is no more, we are extremely clever and exceedingly tame, and we will joyfully put on a uniform and follow a commander's directions, bayonet in hand. This happens as we have language and identities based upon it. These can be founded in religion, ideology, or nationality. Whichever the case, they have deadly consequences. And this instinct to believe in the moral necessity of war is not confined to the infantrymen. It applies to the kings and generals who send them onto the field of battle.

The usual explanation provided by anthropologists for why people seek power is that a position of authority offers us evolutionary advantages and that this lust for power is alike to that displayed by the leaders of bands of gorillas and troops of chimps. That is the essential premise of de Waal's *Chimpanzee Politics: Power and Sex Among Apes*. But even the most cursory study of the biographies of men and women who have pursued the presidency or sought to be kings and generalissimos suggests that this is an inadequate description. Many have been little interested in sexual favors or money. Neither Hitler nor Stalin possessed a harem. They did not acquire great riches either. Both appear to have been genuine and passionate believers in the ideology of their party. Likewise, Francisco Franco was a devout Catholic and a constant husband. Although Queen Elizabeth I was the most powerful woman in England, she never married or had children, and she may have died a virgin. What's more, instead of enriching herself, she used her vast holdings of property to pay for many of the functions of her government, enriching her subjects. A perhaps even more remarkable example of this phenomenon is the Ayatollah Khomeini. Although he was an advocate for a religion that preaches the lawful practice of polygamy and whose founder had many wives, Khomeini is said to have been faithfully attached to one woman for sixty years. Nor did he occupy a palace. His home, Jamaran Hussainiya, was a two-room apartment. Khomeini did not own this flat. Instead, he rented it for $86 per month, and he ruled his nation from a space within it of approximately 120 square feet. Reportedly, he liked it for its bareness and simplicity and its proximity to a mosque at which he preached.

Perhaps even more notable are the men and women who spend their lives as Catholic and Buddhist priests and nuns. The Catholic priests compete with one another for position. Doubtless many aspire to be pope. But in the present day the net result of this will not be a greater number of heirs or an estate they can pass on. So is it not easy to see how it will greatly advance the survival possibilities of more distant relations, adding to what evolutionary biologists refer to as inclusive fitness.

If the purpose of pursuing power is to obtain resources or to produce the largest possible litter, all these folk seem not to have received the memo. In some important way, they were guided by something other than the crude evolutionary impulses observable in chimps and gorillas.

The practice found among early peoples of building giant pyramids is equally revealing. Huge pyramids and ziggurats were constructed during the third and fourth millennium BC in Mesopotamia, Egypt, and Peru. Later, structures of colossal size were built in Mexico, Guatemala, Sudan, Nigeria, and China. All these structures were made of immense stones, and, as most of these ancient people did not make use of the wheel, the pieces had to be transported on sleds and then dragged up to their tops. This consumed much of the wealth and the greater part of the energies of the societies that erected them. Thus, it has become commonplace to suggest that extraterrestrials assisted. A more plausible explanation is that they were the work of masses of slaves. But I think we can exclude the theory that aliens built these monuments, and research on the greatest of the pyramids, those of Giza, suggests that the workers who built it were paid. That is not so difficult to imagine. In some respects, it's almost harder to believe that slaves could have been motivated to do the work while kept in captivity for the long periods required. In either case, these pyramids could not have been constructed unless nearly the whole of the society believed that they were vitally important. We can think of this as a sort of mass hypnosis.

The principal reason that chimps could not have built these structures is not because they are incapable of making tools but because they cannot be so bonded together toward a common aim. Their attention wanders easily, and they pay little heed to the wants, ideas, and interests of others. Nor are they so guided by collective dreams, myths, and fantasies.

Those who worked on these monumental structures were behaving as sheep do when they participate in "sheep tornadoes." These are collective explosions of furious movement in which one sheep follows another in a circle, and the herd winds up spiraling around itself. The flock gains nothing by this, but they believe that their frenetic activity is purposeful. Just as the human doggedly labors on behalf of the pharaoh, striving to help him in his wish to build a pyramid that the ruler imagines will safely deliver him to the underworld, the sheep races ahead behind his leader, convinced that he must do so to reach a place of safety—the very spot he has just been at.

In our own time, we see people of every political stripe declaring their allegiance to ideologies. This is deeply satisfying. For, as we are domesticated creatures, we are insecure and eager for acceptance and praise. What's more, when we adopt the beliefs of those around us it provides us with a ready-made identity. It makes us feel safe and comfortable. Ask yourself: How many movie stars and politicians loudly decrying injustice are motivated above all else by a hunger for attention, respect, or even just a sense of belonging?

Domesticated animals are less intelligent than their wild kinsmen. Wild turkeys and bison are sharper than domesticated versions of the same animals. Wolves have been shown to be smarter than most breeds of dogs. In humans, the process of domestication is one in which we have become exceptionally capable at employing logic in the abstract but inclined toward unquestioning imitation and obedience. There is a lack of self-awareness then in our use of the term "ape" as a verb. A common definition of the word is to "to copy closely but often clumsily and ineptly." While awkwardness in imitation may be an attribute of apes, mimicry and copying are human impulses far more than primate ones. Primates are much more likely to be oblivious or indifferent to the gestures of their peers. We doggedly obey fashion trends. They cannot even follow one another in any sort of line, straight or crooked.

Among the greatest of all proofs of our enormous capacity for obedience appears regularly when men and women meet at an altar: arranged marriage. Historically, more than 130 cultures have practiced it. Indeed, until the twentieth century it probably represented the majority of all unions worldwide as it was normative in China, India, the Caucuses, North Africa, the Near East, and much of Europe.

It is still standard practice in the Middle East and North Africa. Troublingly, it usually takes the form of consanguineous marriage. These are weddings within existing clans. In some countries two-fifths of all marriages are of first cousins. Because repeated in-group marriage can lead to the passing down of harmful genetic mutations, this has damaging effects on the overall health and fitness of the population. Like our habit of suicide, it is an alarming proof of how much our instinctive obedience makes us a danger to ourselves.

Why can humans be kings and presidents without size or strength?

To answer this question, let's return to the behavior of strongmen. One of the more influential kingpins of the moment is Turkish President Recep Erdoğan. Like Khomeini, Erdoğan is a devout Muslim. He is also alike to Khomeini as he is intent on making polygamy normative. To that end, Turkish school textbooks have recently been rewritten and television programs are scripted with the aim of persuading the young to admire a man with more than one wife. Yet, from an evolutionary biologist's viewpoint, Erdoğan is like those leaders we mentioned in the last chapter. He is not benefitting as he has but one wife and four children. In effect, he is promoting an ideology that favors his genes without taking advantage of it. Yet, if this seems contradictory, it is hardly unusual. In the same way, Hitler spoke in favor of allowing multiple wives to decorated German war veterans, though he only married at his deathbed and had no offspring.

Suppose for a moment that a chimp could grasp human speech and that he understood this. That a human might promote selfish ideas without making use of the practices would puzzle him. So why do people act this way? Why can commitment to a cause so override our self-interest—and our libido?

Underlying attachment to our beliefs is a series of other human needs. Among these are the necessity of meaning in our lives and the egoism and pride derived from the sense that we have been steadfast in our devotion to a cause. These help us to create an identity, and, as we are a herd animal, achieving this is an arduous process. Hence, once we have gained this identity, we are loath to give it up, and whether the beliefs we have attached ourselves to are true may be of little significance. The easiest and most persuasive way to demonstrate this is to point to absurd ideas from the past like the practice of trying to determine if a woman was a witch by seeing whether or not she drowned when thrown headfirst into a lake with a weight attached to her. That is too easy though. That is nothing more than a means to flatter ourselves by placing ourselves above our ancestors. To grasp the extent to which people can be hoodwinked by preposterous notions we ought to examine our own follies.

Let us look then at one of the most popular belief systems of the last two centuries: Marxism. Widely accepted on college campuses, it continues to be the faith of many highly schooled individuals. Yet most people who say they are its adherents know little about it, and those who oppose it do so by making reference not to its ideas but to the record of the various communist regimes. But the case against Marxism is simpler. Marxism isn't just a system that does not work in practice. It does not make sense as theory.

Marx believed that the price of goods and services was composed of two parts: the value of the labor on the one hand and the profit generated by it on the other. The latter portion he referred to as surplus value. Marx insisted that over time the amount of surplus value would fall. At the same time, he believed that workers' wages would decline through a process he called immiseration. The reader may notice that these ideas are incompatible as prices cannot be explained by the combination of wages and profits, but both be in decline. If that were so, then prices would have to drop, or one of the two—wages or profits—would need to increase. It defies basic arithmetic.

What would cause this price change? At times Marx attempted to explain this by saying that there would be an overproduction of goods. But wouldn't that mean that prices were actually a function of supply and demand? Seemingly aware of this contradiction in his premises, he did not always suggest that prices would drop. Marx also used alternating concepts of value, sometimes referring to prices as value and at other times insisting that there was something he called "use value." This latter concept is at war with the obvious fact that consumers evaluate their purchases. It also implies that there is some means other than the wants of the purchaser by which to determine what items and services are worth at a given moment. This is a direct contradiction of one of the most basic concepts in economics, what is known as marginal utility.

Marx was consistent in his prediction that rising productivity would cause a lowering of surplus value. But this is obviously contrary to his own premises—and common sense—since if workers' wages were declining and less labor was required to make goods, then corporate profits should be going up, not down.

With very few exceptions present-day economists regard Marxism more as a source of amusement than anything else, and Marxist thought is nowhere mentioned in standard economics textbooks. If you ask an economist about this, they will explain to you that Marx understood neither the cause of prices nor what profits are. Then they will point to the empirical evidence against his beliefs: Throughout his life wages were rising, and working conditions were improving, while technical innovations were making goods cheaper and profits greater. Every one of his predictions was wrong. Marx denied this, and he cherry-picked from official statistics in order to conceal his errors. In some sense, though, that is beyond the point. Marx's economic equations are self-evidently impossible.

These are not the only problems with Marxism considered as a theory. Marxists insist that anyone "oppressed" should be unhappy. When such people say they are fulfilled, Marxists respond by claiming that they are not content. They are said to be deluded about their own condition and suffering from "false consciousness." In this way, Marxists deny the validity of testimony from anyone who disagrees with them. While Marxism is supposed to be a "dialectical" philosophy—one that is strengthened by debate and contradiction—it provides a means by which to reject and ignore statements and opinions in opposition to it. Thus, by its own premises, it renders only the evidence that supports its conclusions as true. At the same time, it claims to be scientific. Yet science is based on a range of evidence, and it should be predictive, though Marxism is neither. In addition, Marxism is not verifiable. Altogether it is a singularly ludicrous and self-referential mishmash of circular, illogical, and contradictory notions. How then is it that at one point almost half the world was ruled by his followers and great numbers of people still believe in it?

If humans regularly employed logic in examining problems, Marx's ideas would have been scorned and forgotten. But what if humans are not practical problem solvers—like chimps—but herd creatures who most often adopt the ideas of those around them? If that is so, then we may begin to get a sense of what is happening. That would explain the building of pyramids. That would explain mass suicides. It would tell us why people become rabidly upset when "their" sports team loses. It would explain fads and cults.

That is also consistent with what we have learned about human patterns of violence. If we are exceptionally dangerous *because* of our natural obedience, then it is of evolutionary value for humans *not* to be reflective about ideas. The path of safety is indiscriminate enthusiasm for prevalent beliefs. Not to believe in what everyone else is saying and thinking is a path to exclusion, sanction—even death.

This also suggests that leaders tend to gain power because they are in tune with their followers. If appeals to emotion, not reason, are the lever they use to obtain power, then the leader will be judged not by his logic but by his seeming earnestness and passion. This is not to say that Hitler, Stalin, and Khomeini weren't capable of impressive amounts of lying and deception. It is rather to suggest that much of their appeal came from their unambiguous devotion to belief systems that they shared with their lieutenants and an inflamed mass of their subjects.

This does not mean that men can't seek to monopolize the mating opportunities in a society. Genetic tests show that sixteen million people in Asia appear to be direct descendants of one figure who lived in the thirteenth century. As this was the age of Genghis Khan's conquests and he had over three hundred wives, it's generally assumed that these are his progeny. The use of eunuchs and the construction of harems among the Islamic caliphates and by various Chinese emperors have had similar if less profuse consequences. It is important to note, though, that most human evolution took place in hunter-gatherer societies, and these tend to be *more* monogamous than present-day Western cultures and *less* tolerant of infidelity and polygamy. While this does not prove Wrangham's assertion that this explains male homosexuality in humans, we cannot deny that it might be a factor in our increasing domesticity. To try to get some sense of why this is and how risky straying can be in a hunter-gatherer tribe, consider that it is likely to be:

1. Composed of fewer than one hundred people. Most hunter-gatherer societies are this small.
2. Made up of people who spend much of their time gossiping and talking about the members of the tribe.
3. Formed of related individuals.
4. Likely to have roughly the same number of adult women as men since men lost in hunting accidents and by violence will tend to balance out against women who die in childbirth.
5. Constantly on the move and often struggling to find food. Thus, its members have little time or tolerance for acts that spur jealousy or conflict within the tribe.
6. Possessed of few or no slaves. In fact, hunter-gatherer societies often adopt those captured in war, accepting them into the tribe.

Through this list, I think you can see an additional reason why we might have evolved to value privacy in sexual relations and why nearly all nations and peoples throughout history have adopted dress that covers the genitalia. These are ways to deter fighting. And, while humans are not the homicidal maniacs that chimps are, trying to seduce another person's wife or husband is a good way to get yourself thrashed or killed. That eliminates your genes from the gene pool. This means that most men aren't likely to be focused on seducing other men's wives.

Hence, powerful leaders tend to have goals and ambitions that are of much greater interest and concern to them than maximizing their chances of reproduction as a chimp and gorilla might wish to do. Napoleon serves as an example. The French emperor is often called a "son of the revolution," but his real belief and fascination was with himself. Yet, while he had a strong libido and was amoral, he only fathered one legitimate heir, and DNA evidence suggests that only two of the illegitimate children attributed to him were his.

In our own time, a remarkable number of men and women have created great fortunes and then agreed to donate most of their accumulated wealth not to their children but to philanthropy. According to Business Insider, more than 211 American billionaires have pledged to give more than half their riches to charity. Nor is this new. One has only to examine the names of some of our leading universities to notice the pattern: Vanderbilt, Duke, Stanford, Colgate, Carnegie-Mellon. Doubtless most of these men were driven, greedy, and unprincipled in their business lives. They wanted to make huge sums, and they devoted themselves to quashing competitors. What prompts them then to so freely part with their fortunes, leaving their heirs with vastly less? While this is obviously partly motivated by altruism and partly by vanity, there is something more. The "Giving Pledge" that Bill Gates and Warren Buffett put forward, which snared those 211 billionaires, is a form of social pressure, and this affects the super-rich as much as any other category of individuals. Thus, we again see how social humans are: Even the superrich cannot free themselves from the urges and compulsions that make us human.

One further ironic demonstration of this phenomenon by which seeming evolutionary "imperatives" turn out not to be may be found in the mating habits of the recent crop of eminent evolutionary biologists. Few born since the Great Depression have spawned more than two children. Richard Dawkins, for example, has been married three times, but he has fathered just one child. Jane Goodall is twice married with one son. E. O. Wilson was wedded once, and his marriage lasted for sixty-six years. He and his wife had just one child. Prolific as they are not, an extrapolation of present trends suggests they will go extinct in a few thousand years. In more than a few cases, they will do so insisting that humans are creatures who rationally go about seeking the means by which to produce the greatest possible number of surviving offspring.

CHAPTER 10

MAN, THE UNTHINKING ANIMAL

ONE OF THE MOST COMMON expressions of people over the age of forty is: "It's what everyone was doing at the time." This is how we explain our youth. It is an implicit admission of an undoubted truth: In considerable degree we are herd animals.

One of the most obvious proofs of this turns up in the countless advertisements we are presented with. These campaigns are the work of professionals who excel at hooking customers, selling us on things we haven't previously desired or even contemplated purchasing. Among their most common pitches are those telling us that the products we should buy are the most popular. A movie, we are informed, was the top motion picture at the box office. A cola is drunk by more than any other. No further explanation is required.

We see this impulse toward the herd mentality most clearly in one industry: fashion. Permit me to ask readers to think about that in terms of the plot and story of the recent movie *The Devil Wears Prada*.

Before we do so, however, I should admit something: My taste in movies and plays is typical of the herd. This is sufficiently true that in the course of several years in which I served as a movie and drama critic, one who reviewed something like a hundred different plays and films, I only disagreed with the bulk of the audience on a few occasions.

One instance in which I can remember being at odds with the general view is with the film *The Devil Wears Prada*. It has positive online ratings from both audiences and critics. Permit me to explain the story, for those who have not seen it. A young woman who is bright and intellectual but not especially stylish (Anne Hathaway) winds up as the assistant to a tyrannical fashion magazine editor (Meryl Streep). Her boss continually harangues and bullies her. In the midst of this, she begins an affair with a handsome potential rival (Simon Baker) to her employer. Recognizing the aide's situation, he asks her to help him in his plot to unseat her vindictive boss, offering to promote her if she will do so. Yet, out of loyalty to the woman who has behaved so shabbily toward her, the assistant turns him down.

One reason I didn't like the movie was that I couldn't figure out why the assistant's devotion to her employer was supposed to be commendable. Nor to my mind did it make clear what's so admirable about fashion. The best it could do on that front was a speech in which Streep informs Hathaway that the color of the sweater she is wearing reflects a chain of previous fashion trends. As Streep puts it, "What you don't know is that sweater is not just blue. It's not turquoise. It's not lapis. It's actually cerulean. And you're also blithely unaware of the fact that in 2002 Oscar de la Renta did a collection of cerulean gowns. And then I think it was Yves Saint-Laurent…wasn't it? Who showed cerulean military jackets?" Streep's point is not that the choice of color makes the clothes more beautiful but rather that it's possible to manipulate the public toward arbitrary choices. We are supposed to think that all this is wonderful—so long as we are among the favored few who become aware of these trends before the benighted masses. How this appeal to snobbishness is supposed to be admirable left me a little confused.

That said, I don't mean to suggest that clothes can't be gorgeous or that concern for them isn't a worthwhile activity. Undeniably, the world is a more appealing place when it is populated by people with taste in clothes and an amount of style. My objection is that this has little to do with what Streep is saying. Indeed, as many writers have noted, the most famous and notable embodiments of elegance—like Cary Grant or Diane Keaton—paid little heed to the latest trends in fashion. Instead, they developed a personal style that reflected their own taste.

Both in the decision of the character played by Anne Hathaway to protect her malicious employer and in its promotion of the fashion industry the movie was celebrating the notion that people should be subservient. It was endorsing the idea that we should obediently follow the rich and powerful. In some sense, this stands to reason. For where clothes and manners are concerned, the fight to free oneself from the herd is one most of us will not dare to take on. As Voltaire explained it in the form of a riddle: "What do all wise men ridicule and obey? Fashion." Abiding as our herd instincts are, they are hard to resist. When no one is wearing white socks with dress shoes, you are bound to draw unwanted attention if you do, and, as the British essayist G. K. Chesterton observed, "It's more withering to accuse a man of bad taste than bad morals." Standing out for faulty attire can get you kicked out of good society.

Yet herd mentality powerfully displays itself in many human activities. This includes sports, politics, art, architecture, music, and the actions of financial markets. Both sports contests and political parties awaken our attachment to the group, and this plays some role in the worshipful regard accorded to sports stars and politicians. This devotion appears even as our connection to the herd clouds our judgment about the prospects of our preferred teams and the character of our leaders.

A team of Princeton and Columbia University researchers interested in herding effects in musical tastes found that when people were permitted to sample a batch of unknown songs that they showed a marked preference for whichever songs they believed—rightly or wrongly—were most selected and listened to by others in the test pool. This affected both their ranking of the songs and which they themselves listened to.

Similarly, while we consciously dress to conform, we are usually not even aware that our thinking about currencies, stocks, or politicians has been distorted by groupthink. This impulse to adopt the latest trendy notions—to believe what is momentarily au courant—may be even more insinuating and commanding because we do not realize that our attraction to these opinions has been guided by transient forces.

Financial markets are distorted by two emotions: fear and greed. Fear drives markets below rational levels and greed above them. Stimulating each is herd mentality. Greed drove the tulipmania of the seventeenth century. At its height, tulip bulbs were worth more than Dutch houses. More recently, the Japanese property boom of the 1980s culminated in a bubble in which the value of land in Tokyo exceeded that of the entire United States. The best investors know that markets are guided by herd instincts. Thus, George Soros is reputed to have said that he does not like to buy stocks or currencies until their market is already going up. Soros knows that he cannot fight it. Employing the same logic, the economist John Maynard Keynes, who was a famously shrewd stock picker, liked to advise clients not to base their judgments on rational analysis alone because, as he put it, "the market can remain irrational longer than you can remain solvent."

Fascism and communism are political systems whose express intention is to create a unified mass. Fascism holds that humans are sticks that can be bound together. The bundle is supposed to be stronger than the single rod. What is most important is not the individual but the group: the herd. And while we may be inclined to think of herds as harmless, they are anything but. Zebras, cattle, elephants, reindeer, wildebeests, walruses, wild horses, and rhinoceroses all stampede those in front of them, and pigs will not only knock people down but eat them. Although herds can be taught to be peaceful, they can also be trained to be aggressive, and in their numbers they are inherently dangerous. Humans can stampede one another, of course, and in our time we have seen deadly instances of this during fires, floods, and even store sales.

Just as a deer is endangered by falling out from the herd, we, as humans, are at risk by ourselves. If we are not part of a herd already, we want to join one. Outcasts do not seek to be outcasts. This problem is especially acute for the young and those lacking in a sense of identity. This is shown in fraternity hazing rituals. Like the Milgram experimenters, the older members of Greek organizations convince themselves that the cruelty of their induction procedures is a form of character building and a necessary part of the process of bonding. They are *helping* the pledges. In any event, who wants to be thrown out of the group for not going along with what the person in charge suggests?

This desire to fit in and to find a place provides us with a partial explanation for why we are susceptible to recruitment into cults and crackpot ideological movements and why organized religion can be protective for teenagers. Those without a highly developed sense of who they are provide fertile soil for religious charlatans and charismatic extremists.

The fraternity hazing ritual points us to another truth about humans that distinguishes us from chimps and gorillas. When chimps and gorillas form alliances, they do so in order to arrange the group's hierarchy and to ensure themselves access to the distribution of resources. While that is certainly true in human society, the creation of our intra-sex bonds plays an essential role in the formation of our identity. Intra-sex bonding is a means by which we establish our sense of self, and for this reason we have enormously complex rites and rituals associated with it. Many of these are as singular and as invested with emotion as those connected with marriage. And since those most in need of an identity will be the most drawn to these groups and to their rites, they have particular appeal among young people lacking in parenting. Thus it is that teenage gang members, many of whom have been without fathers, go to enormous lengths in their initiation rituals. So, too, the children of the rich, feeling ashamed and alienated from their environment and often raised by servants or indifferent step-parents, serve as especially ripe targets for offbeat groups with outlandish beliefs. Hence, one of the common observations about 1960s radicals was that they were born into privilege. Among this number were publishing heiress Patty Hearst, movie actress Jane Fonda, and nearly all the leaders of the Weathermen terror faction.

There's something else that should be pointed out here. We tend to equate objectivity with dispassion. For that reason, we may be inclined to think that a tame creature is more rational. We see this idea in the character of Spock in the *Star Trek* TV series and movies. Spock is completely rational because he lacks for a human's normal allotment of feelings. This premise finds its corollary in the contrast between the extreme rationality encountered among Asperger's sufferers and the hyperemotionalism and irrationality seen in adolescents. In many respects, though, this rule works in opposite fashion when applied to a species. Chimps tend to be violent and temperamental. However, when acting under the influence of the ideology of the group, it is humans who are apt to be herd-like in their thinking and, therefore, irrational, not chimps. The more tractable an animal is, the more it may display faddishness and lack of reason in its behavior as a member of a herd.

In a curious way, this brings up our propensity for creating systems of monetary exchange. The existence of money may be the greatest of all proofs of our innate obedience. While money serves a function for us, this is only because we implicitly and unthinkingly accept its meaning. Were we not tame and courteous, money would have no value.

It is notable that amidst what were supposed to be the most radical anti-capitalist revolutions of modern times, only one attempted to do away with money. That was in Cambodia under Pol Pot. By contrast, even during Mao's Cultural Revolution, printed money continued to circulate. That was in spite of the fact that those leading this campaign said they violently objected to the unwelcome consequences of monetary relations. It seems that in their hearts Mao and his various followers remained attached to the belief we should all acknowledge that a bill or coin ought to mean the same thing to one person as it does to another.

Not surprisingly then, many rebellions have been spurred by outbreaks of inflation. For inflation not only robs savers, it also strikes at our intuitive belief that the ordinary action of paying someone for their goods or services should follow a basic, if unstated, principle of courtesy and fairness, one in which we collectively agree to the fiction that units of currency represent something other than what they actually are: scraps of paper, arbitrarily assigned with meaning and significance.

There is an obvious contradiction, of course, to the claim that humans are empathetic and altruistic: We are the only species that enslaves its own kind. Moreover, we subjugate other animals, and human wealth and activity have depended upon this.

The use of domesticated animals can be alike to slavery. Still, much livestock raising is done with a devotion to animal welfare. For that reason, domesticated animals routinely outlive their wild relatives, and some pass through lives of ease and comfort. That is not always the case though. Modern industrial hog farms cut off pigs' tails because they are so crowded and stressful that otherwise the pigs would bite off one another's tails. Cows are forcibly impregnated, then provided with corn to fatten them up, no matter that it is a food they can barely digest. That makes them so sick that they are "downed": They cannot walk and instead lie on their side.

Slavery and animal husbandry combine themselves in the practice of gelding. Early on in the practice of livestock raising, humans discovered that males of a species can be made more docile by cutting off their testicles. Eventually, this knowledge was employed upon people. And, like so many abominable practices, it found its apologists. Thus, as an admirer of the Persian rulers, the ancient Greek writer Xenophon specifically argued for the employment of eunuchs within the Achaemenid empire.

Among the most famous of the surviving full-length defenses of the practice is one from Theophylactus of Ochrid. A figure of the eleventh-century Bulgarian court, he served as an archbishop in the Eastern Orthodox Church, and his scriptural commentaries were so admired that Thomas Aquinas placed some in a collection he prepared. Defenses of eunuchs also can be found in ancient Hindu tradition through the writings of the members of the ancient sect of Vashnaism, which devoted itself to Vishnu and his avatars.

Castration was one of the more egregious aspects of slavery not seen in the United States. While some readers may not know it, the slavery practiced in the Confederacy was among the less cruel versions practiced during the nineteenth century. In fact, it was only in the Southern states that the number of slaves grew by natural increase. In every other part of the New World, slaves were worked or beaten to death, or they were allowed to fall in colossal numbers to tropical diseases. Then they were replaced by new multitudes brought from Africa. By contrast, in the United States the high monetary value of the slaves prompted slave owners and overseers to make an effort to attend to their slaves' health. Yet the barbarity remained. When a slave arrived at an auction, he was stripped naked, and his body cavities were searched. As photographs show, the "peculiar institution" that Southerners so eagerly defended required lashings, and slave owners regularly separated families by sale and purchase.

The arguments advanced for slavery and castration should point us to something distinctively human: Although we are more innately empathetic than chimps and gorillas, our capacity for groupthink can further our selfishness and make us equally oblivious or indifferent to suffering. This follows logically if we conceive of ourselves as a species that has a strong innate impulse to adopt received ideas and a resistance to stepping out on our own and thinking for ourselves. Research on brain response has shown that social rejection causes stimulation of the same parts of the brain—including the cingulate gyrus—activated by bodily discomfort. In humans there is actually a link connecting feelings of isolation and physical pain. Hence, as thinking for ourselves makes us lonely outsiders it can not only be excruciating emotionally, but it can induce something that feels like the opening of a wound. That provides plentiful incentive to accept or defend the prevailing ideology.

CHAPTER 11

WHY WE CAN RUN MARATHONS BUT NOT OUTRUN BEARS

AMONG THE MOST COMMON MISCONCEPTIONS about human evolution is that we are descended from chimps. Rather, we have a common ancestor with chimps. What precisely that predecessor was like we cannot be sure. Richard Wrangham has said that our forebear was likely much closer to a chimp than a human. This is like saying that a painting of the Crucifixion might have religious themes. The statement tells us much more about the field of primate anthropology and its assumptions than it does about our predecessor. It's a given that this is so. Our ancestor must have had the characteristics of primates. How could he not? He was a tree-dwelling herbivore who rarely traveled far from his home. Logically, it follows then that he was territorial and not especially tame. We also know that his brain was far smaller and his intelligence was relatively unremarkable—certainly when compared with creatures like whales, dolphins, horses, and elephants. Yet it's undeniable that we don't know his precise nature, long gone as he is.

Awakening of interest in our kinship with apes has taken place over the last century as more and more has been learned about their patterns of behavior. The degree of our ignorance prior to that time is startling. In fact, until the 1930s, it was thought that chimps were strict herbivores. Then researchers discovered that they sometimes kill and eat colobus monkeys, and monkey meat can play a role in determining the attractiveness and desirability of chimp males among the females they seek.

Of course, we are still limited in what we know about present-day apes in a way that we are not held back with respect, for instance, to felids and canids. After all, while a full-grown lion may seem like a fearsome creature, he is capable of being semi-domesticated, and trainers can and do spend great amounts of time with them on warm and affectionate terms. In the same way, we live with German Shepherds and Doberman Pinschers, which are quite capable of tearing us to shreds and eating the scraps left behind; we cohabit with dogs that have been trained to excel at killing us, when and if they are commanded to do so.

As we have said, chimps are different. Early on in her research at Gombe Stream National Park, Jane Goodall managed to give a few chimps bananas right from her own hand. But she no longer does. Nor will she permit her aides to do this. Goodall's reasons for prohibiting such actions are not guided simply by safety concerns. These are increasing though. Determined to protect their narrow areas for foraging leaves and stems, chimps attack competing gorillas, kill them, and leave them, mostly uneaten, for consumption by scavengers.

That chimps and gorillas must be kept at a remove is a problem that applies as well to research done on apes kept in zoos and primate research centers. In de Waal's words:

> None of us dares to go in among the chimpanzees. Their keeper and I are on very friendly terms with some of them, but only when they are in their sleeping quarters, and there are bars between us. Zoos make it a rule never to trust any adult chimpanzee fully. They are no heavier than humans, but they are *much* stronger. The problem with chimpanzees in a zoo is that they are well aware of their superior strength. This plus their temperamental nature makes them deadly. [Emphasis in the original.]

There is a reason that there are lion tamers and dolphin, seal, and dog trainers, but no chimp tamers.

This leads us to a specific issue with de Waal's claims. His book *Chimpanzee Politics: Power and Sex among Apes* was based on his observations of captive chimps, and many of his subsequent assertions about bonobos was founded in work with captive bonobos. Yet it is an axiom among experts on animal behavior that creatures in captivity behave differently than those in the wild.[6] Life in a zoo affects "mate choice, courtship, mating and rearing of offspring." Over time it can also induce genetic changes in a population. These can occur in a single generation. This means that not only do we have to question de Waal's general conclusions about apes and their relations to humans but the underlying methodology of his research.

Nonetheless, his research is consistent with everything else we know about chimps and gorillas, and this brings us back to one of the main reasons we have identified for why chimps are not much like people: their lack of cooperativeness and sociability. To understand the implications of this fully it may be worthwhile to consider one example of *human* behavior. If you live in a small town in Iowa and someone's house burns down, you may be called upon to assist in a joint effort to build him another one. This involves the whole community in everything from hammering boards and laying a level foundation to baking pies and cobblers.

Chimp troops are small, even much smaller than Iowa farm communities. Yet if a mother chimp's nest is destroyed by a thunderstorm, the troop won't help her rebuild it. Even the possible fathers of her children won't assist her. By contrast, nearly all commercial activity depends upon our cooperation. As a well-known economist once pointed out, there is not one person on the planet who knows how to make an ordinary pencil. Instead, many people coordinate their activities to obtain the graphite, the material for the eraser, and so on.

---

6   This even applies to humans. When people are enslaved, we breed poorly. See Jane Spencer, *Writing About Animals in the Age of Revolution* (Oxford, UK: Oxford University Press, 2020), 164.

Should you think my example of the Iowa farm community is unfair and one-sided as a comparison between humans and apes, then consider a group of people who live right alongside the chimps in the equatorial forests of the Congo: the Mbuti tribesmen. As the Mbuti are pygmies, they are little people. In fact, average height for a male tribesman is four feet nine inches. Thus, they are significantly smaller than chimps, to say nothing of gorillas. But they, not the apes, are the big-game hunters. In fact, the Mbuti are among the last true hunter-gatherers in the world.

During the rainy season, the Mbuti live in huts at the edge of the forest. Within each resides a nuclear family. But, once the rainy season passes, the tribe moves back into the heart of the forest. They go there in search of game. However, the Mbuti do not attack chimps or gorillas. Because they are animists and their mythology holds that the chimps are vital spirits of the forest, they believe that the chimps must not be harmed.

This does not necessarily show innate goodness on the part of the Mbuti since they have little reason to attack the chimps. That's as they are not reliant upon the leaves and stems that chimps and gorillas depend upon as their sources of nourishment. Instead, the Mbuti spend most of their time foraging for legumes, beans, and yams; fishing in streams; and hunting for wild pigs and okapis. The okapis—a species of giraffe—are much too fast for chimps. But the Mbuti have bows and arrows.

Women do not participate in this hunt. However, Mbuti women are eager participants when the okapi are to be captured with nets. This requires immense cooperative skill. Women and even children will try to corral the okapis, forcing them into the snares. This requires everyone to work together, alternately talking and plotting and making use of stealth and silence. Some of the acquired bushmeat is then traded with other tribes in the region who provide the Mbuti with pots, jars, baskets, and other items of value in return.

In all these activities, the Mbuti are below the chimps, clustered in the trees above their heads. Thus, the title of Jane Goodall's famous book about chimp life, *In the Shadow of Man*, reverses matters, placing the monogamous hunter above the polygynous herbivore. As we are bipeds with feet made for walking, we are in their shadow, surviving because of our capacity to work together.

In this way, we find once more that we are alike to herd and pack animals—creatures like horses, dogs, camels, elephants, and hippos. Like them, we are engaged in activities that require patience and endurance. The second of these two attributes distinguishes us not only from primates but most carnivores. They tend, after all, to be sprinters, not distance runners. This persistence and the capacity for prolonged pursuit are an additional means by which we compensate for our lack of claws, big teeth, agility, and speed.

So what other animals are endurance runners? One analysis of the best "marathoner" species came up with this top six: horses, humans, camels, sled dogs, pronghorn antelopes, and ostriches. Another placed us just behind the dolphins and whales, setting those cetacean species at the head of the list of animals with the greatest stamina.

The biological mechanisms underlying our capacity for prolonged exertion are only coming to be understood. One puzzle piece is our unusual skill at sweating. That permits us to relieve ourselves of excess body heat. Humans can perspire about ten times as much as chimps. No primates sweat much. The animals that evolved to rely upon perspiration are herd animals like horses and zebras. This assists them as it does us. We can continue running or walking when other creatures would have to stop for fear of overheating themselves.

Another means by which we have become a "marathoner" involves a transformation of our hearts. Within the last few years researchers discovered that they aren't just metaphorically distinct from those of chimps and gorillas. They are literally different. Since chimps and gorillas don't need much in the way of physical endurance but they do engage in sudden bursts of rapid motion, there are occasional moments when their blood pressure spikes. This happens during many of their routine activities, say after they climb a tree or fight off a foe. Common as those demanding tasks are for chimps and gorillas, ape hearts have thick walls and a round shape. The sturdy walls and the compact design help them when their blood pressure abruptly goes up. This prevents a heart attack. However, it does not yield as much sustained blood-pumping power as a human heart. That's because the large left ventricles of our hearts continue working for long stretches and they are bigger overall.

Not only do we have larger hearts, but we mature later. Humans normally reach puberty between ten and thirteen. Dolphins attain it between eight and thirteen. Although they usually cannot conceive at the age they begin ovulating, chimps become sexually active between five and eight, and on average chimps in the wild do not live past age fifteen. This is another reason why humans require bigger brains and greater patience and skill at planning. We were designed for longer life.

Paradoxically, that required us to look less adult and mature. Humans have been taking on degrees of paedomorphism. This is the phenomenon by which domesticated animals come to look more like their child and adolescent forms. Research on the domestication of foxes shows that this is an evolutionary side effect of selection for tameness. As creatures become less hostile and aggressive, their eyes get bigger, their bodies become slimmer, and their growth is slower.

We display evidence of this. Our eyes are larger than those of chimps and gorillas. Our teeth come in less quickly. We are less muscled. We take longer to reach full size and strength. This contrasts humans not only with chimps but Neanderthals, whose adult teeth appear to have been in place at earlier ages and whose torsos were stockier and more powerful.

As primatologists operate with the assumption that humans are primates—though we lack for one of the two defining traits of the order—they had long believed that our basic metabolic rate would be roughly the same as that of chimps, gorillas, and orangutans. And as we grow more slowly than apes, it stood to reason that if there were a difference, that we would be the species with a slower metabolism. But this is not what researchers found. In fact, our basic metabolic rate is much higher. Adjusted for body weight, we use 20 percent more fuel each day than chimps and bonobos, 40 percent more than gorillas, and 60 percent more than orangutans. As a noted researcher recently observed, apes seem like sloths when compared to humans. Moreover, these differences are not a consequence of how much and how often we are on our feet pacing about. It turns out, in fact, that the people of East Africa who walk many miles each day have no higher metabolic rates than sedentary Americans. It seems, instead, that we have been programmed by evolution to use many more calories. This serves to assist the working of our bigger brains, and it's made possible by our more nutrient-dense meso-carnivorous diet.

That conclusion is consistent with another research finding. Scientists recently learned that there is a general pattern by which carnivores tend to have higher metabolic rates and more caloric consumption than herbivores. This again raises the question of whether we are really well-suited to being called primates: creatures reliant on a plant-based diet.

CHAPTER 12

SENSE, SENSIBILITY, AND SPOKEN LANGUAGE

THE MORE SOCIABLE A CREATURE is, the more it is a herd creature. This rule applies not only across a range of species but within one. People who are particularly social are those most likely to take on the protective coloration of fashionable ideas. Think of a spectrum of occupations, extending from magazine editor, politician, and actor on one side, to medical examiner, lighthouse keeper, and accountant. The more social the job category, the more likely it is that the person holding it is to be drawn to au courant thoughts, actions, and ideas.

This has an implication that may not be immediately apparent. Those employed in positions as "influencers" are those least likely to be independent thinkers. Since journalists tend to be highly social, this means that they also tend to be especially *disinclined* to think for themselves. This is part and parcel with George Orwell's oft-quoted remark that "One has to belong to the intelligentsia to believe things like that: no ordinary man could be such a fool." The idea was expressed in another fashion by the radical journalist Alexander Cockburn who observed that "the first law of Journalism [is]: confirm existing prejudice rather than contradict it." Ironically, both men exemplify this as each was committed to some version of Marxist dogma. For Orwell this was the Trotskyite variety. For Cockburn it was Stalinism.

The usual explanation for this quite recognizable pattern in which learned people are prone to being duped by concepts like Marxism or deconstruction is offered in their divorce from the problems of everyday life. It may be just as much a matter of self-selection. If they were really of a naturally independent mind, it's probable that reporters and congressmen would have gone into different lines of work, as for example pharmacologists or actuaries, and they would not be so intent on receiving invites to nominating conventions and newspapermen's after-hours rounds of drinks. That the opinions of scientists are more diverse and less predictable than those of journalists and elected officials is partly a function of the greater sociability of reporters and aspirants for office.

In an earlier chapter, I mentioned a study showing that people's opinions of songs could be manipulated by falsely guiding them to differing beliefs of what others thought of the music they were listening to. But average people still had some consciousness of what a very good and a bad song was. Only intellectuals ever took to attending concerts where atonal music was played and then, after hearing such painful noise, stood up and applauded. Uneducated people are often harder to fool than intellectuals.

One of the crucial differences between herd animals and primates is shown in how hierarchies are constructed. As we have noted, chimp and gorilla hierarchies are largely determined by the brute strength and aggressiveness of the leaders. Insofar as herd animals have hierarchies, they are based on social awareness and social skill. This is another proof, of course, that we are not primates. But it carries a further implication. Researchers who were interested in seeing which goats were best at investigating and solving a problem—finding food that had been hidden—observed that the ones who excelled at this tended to be worse at relating to other goats, and they ranked lower within their herd. They were the "outsiders." The researchers concluded that this was a general pattern among herd animals.

This has significant implications. These apply to science and philosophy. While scientists may be less susceptible to the impulse toward "groupthink," it is a profound problem within every field of scientific research, and in academia it may be exacerbated by the tenure process. By requiring college professors to get a broad base of support from senior faculty before they can be assured of their job security, we discourage independent thinking, and we may be selecting those least capable of it. We are seeing the consequences of this in our own time with the adoption of academic speech codes. Ironically, a system that was implemented with the aim of supporting independent thinking on university campuses is producing the opposite effect.

We can also see this in the frequency with which "experts" make wildly erroneous claims. During the 1970s it was common to hear psychiatrists insist that cocaine was not addictive, and in the 1990s medical professionals assured the general public that it was perfectly safe to prescribe powerful timed-release narcotics. Because certification in these fields requires repeated steps through which one is inducted and approved, this deters questioning, and it promotes intellectual passivity. The more elaborate the procedures we adopt for accreditation in a subject, the less we can expect originality within it and the personalities equipped to provide this.

This phenomenon was recognized by the American sociologist David Riesman, who referred to it in his book *The Lonely Crowd*. In his terminology, this pattern involves the replacement of the "outer-directed" personality for the "inner-directed" one. The person of an independent character gains less favor and success than the one who is manifestly and intentionally a product of his place, class, and time. Outer-directed personalities seek to conform to social norms, they aggressively seek approval, and they court popularity. A society led by such people venerates fame and celebrity. Yet because these are innate human desires, the yearning for status and adulation affects all human societies.

What does that mean in terms of politics and political discussions?

One of the major disputes in philosophy revolves around the degree to which we can rely on our own powers of reasoning and the extent to which these should be placed above established laws and traditions. Admirers of Edmund Burke tend to disparage "pure" logic, arguing that an undue confidence in our own understanding is hazardous. Burke passionately believed in limits upon the role and authority of the state, and he was convinced that radical attempts to remake the government and society were damaging to both.

Consequently, his followers have seen in his writings an argument on behalf of long-standing traditions and of systems of checks and balances. They conceive of these as a means to restrain the impulses of politicians and to deter the momentary wishes of the mob, and if humans are inclined to collective delusion, then surely this makes some sense as it stops us from following politicians in the way that sheep are capable of following their most socially adept leaders: right over a cliff.

That those most drawn to intellectual and political leadership are the least able to think independently may in considerable measure explain events like the Blackshirt movement in Italy and the Cultural Revolution in China. These were guided and promoted by members of the intelligentsia. Mussolini was a newspaper editor before he was a politician, Mao was a full-time agitator, and Chou En-lai was a night-club performer. Similarly, Hitler was a painter. None of these men worked at lonely, humdrum jobs. All were attracted to the life of the social personality and were eager for fame and adulation. Nor is this description limited to these four. Their most ardent disciples likewise composed themselves from among the artists, journalists, and office seekers of their day. Should it surprise us then that they were drawn to beliefs that displayed trendiness more than common sense?

Sociability is a path to safety. It protects the pig, deer, dolphin, and human from attack by predators and rival bands. In times of chaos and strife, this becomes especially desirable. Then we are particularly obliged to pay attention to what other people think, anticipating sudden shifts in the direction of the herd. This is easiest for a nature that is attracted to the fashionable notion. It requires keen listening skills and a certain instinctive obedience. This applies to the leader as well as the follower, and it is a characteristic aspect of our species as it is of dogs. However, it is much less typical of chimps as they are lacking in our facility for listening, observing, and remembering what they have seen and heard.

That is not conjecture. One study found that dogs were able to sustain their attention for a simple task and remember what they had learned for two minutes. Chimps managed only twenty seconds. A canine's concentration was at least six times better. In one study chimps performed the worst of twenty-four mammal and bird species studied, only outperforming bees.

That might be part of the reason why chimps have trouble completing simple tasks that dogs can perform. This pattern was noted by researchers at Germany's Max Planck Institute. The institute's Dr. Katharina Kirchhofer took 23 chimps and 32 dogs into a lab and asked them to retrieve a pair of objects that she directed them to by pointing. If they performed the task, they were given a treat as a reward. Nine of the dogs succeeded. None of the chimps did. Subsequent research at the University of Arizona offered further evidence for this. Working with larger study groups—552 dogs, 105 human toddlers, and 106 chimps—the American researchers found that human toddlers and dogs performed comparably well at following gestures and completing guided tasks that required collaboration. The chimps did far worse.

Not only do chimps do poorly when compared with dogs on tests that require concentration and focus, they frequently perform worse than other primates. This is a point that Frans de Waal made. He attributed this to the fact that they "lose interest once they have figured it [the test] out. Compared to a pair of rhesus monkeys tested in the same lab, our chimps underachieved, which goes to show that performance and intelligence are not the same thing."[7]

This finding matches with other work done by one of the best-known of de Waal's colleagues in the field of primate research: Masako Myowa-Yamakoshi. In 2015, she and two of her collaborators released a study they had undertaken to look at how chimps and humans responded to videos showing someone engaged in what they termed a "goal-directed action." As they put it:

---

7 De Waal seemed to have missed the point: that in this case performance and intelligence are the same thing; that for the purposes of the test he was giving there was no demonstrable intelligence without sustained powers of concentration.

Human adults and children attend to the actor's face during action sequences, and this tendency is particularly pronounced in adults when observing that the predicted goal is not achieved. Chimpanzees rarely attend to the actor's face during the goal-directed action, regardless of whether the predicted action goal is achieved or not.

Chimps weren't much interested in *why* actions are performed. They were curious about the outcome.

Yet we should not be entirely gratified by these results. After all, the instinctive readiness of humans to listen and obey provide us with the skills which we employ in the formation of brigades of fascist henchmen and companies of Red Guards.

That human sociability leads to added complexity in our relationships does not negate the fact that we can rightly talk about personality in animals. While the term "personality" is grammatically peculiar, anyone who has had a pet knows how apt it is. There is a difference though between personality and sensibility. Chimps, for example, may display widely varying personalities and temperaments. Yet the more refined concept of sensibility does not so easily apply. That implies a level of taste and awareness that a creature with language has, like a human or a dolphin.

I once had a chance encounter with the well-known psychologist Carol Gilligan. Introduced at a party, we wound up speaking for the better part of two hours. As my parents were psychologists, I was familiar with her work, and our conversation focused on the disregard that she and my father shared for B. F. Skinner's theories of behaviorism. This matched to their mutual belief that rats did not serve as good models for human psychology. Toward the end of our chat, I mentioned a theory of mine about human relationships. This was that the couples who got along the best had different temperaments but a shared sensibility. Gilligan pondered this for a moment and then said that it described her relationship with her husband. Many people have since told me the same thing about their marriages. This concept cannot be applied to chimps, though, since they don't pair-bond, and they haven't anything we might call refined sensibilities. There never yet has been a chimp couple that debated their respective taste in films, or even in tropical fruits and tree nuts. How could there be? They have neither couples nor words.

That gets us to another point. Spoken language isn't only a means of communication but also of interpretation and perception. If it is true that we cannot understand a bat without experiencing the process of echolocation, then it follows that we cannot fully grasp the thought of an animal that lacks words.

In John Locke's *Essay Concerning Human Understanding* he observed that those speaking a language without a word—say for stabbing—may be hard-pressed to appreciate the concept. If this is true of a single word or idea, think how powerfully it is so for a creature devoid of spoken language. This is not an animal that thinks or sees the world as we do. Even our application of the metaphorical word "see" in the last sentence must be lost on him.

If it is impossible to understand what is distinctive about people without first acknowledging the centrality of spoken language to our experience of life, then we must ask which other creatures possess it.

Many animals have recognizable warning signals that they relay to their peers. These calls can distinguish different types of threats. Chimps possess this ability, but so do meerkats, prairie dogs, chickens, blackbirds, mustached bats, and countless other species. In like fashion, bees have evolved specific dances that convey warning messages, and squid change color to caution their fellow cephalopods about the approach of predators. There is even evidence of communication within some plant species.

None of this is what we mean when we discuss spoken language. Spoken language is composed of great numbers of distinct sounds that represent specific ideas and objects. We grammatically arrange these and combine them to produce recognizable meanings through the process that we call syntax. By this standard, the animals that appear to have the most complex forms of spoken communication are whales, dolphins, and elephants.

Given the centrality of language to human existence, let's make a particular return to the question of how humans and dolphins are alike. It should be no surprise that we have learned that dolphins can communicate critical information to one another about breeding, hunting, and much else. Nor should it be any surprise that they are empathetic and social. Dolphins communicate with one another in multiple ways. Because they are sensitive to shifts in the water, they can "tap" one another, making a gesture that causes a ripple. This will be felt by another dolphin in the same way that actual touching would.

We can connect this to an interesting finding of researchers on cetacean neurology. They have discovered that many dolphin species have an extra layer of brain matter, what has been called a paralimbic system. This seems to be a section of the brain that processes emotion. The implication is that dolphins have one more area for feeling and sensitivity, an entire region absent in people. It seems that their evolution may have made them even more emotionally aware than we are. That makes sense given their use of spoken language, and it appears in two forms. They use whistles to indicate their name and their location to other dolphins. Beyond this, they have a system of communication based on high-pitched clicks.

Denise Herzing is considered the foremost expert on the subject of dolphin language. She led the team of cetacean researchers who first demonstrated that Atlantic spotted dolphins are named and that their name reflects their matrilineal descent. Like most scientists, she is cautious in making claims about the nature of their language. She does point out, however, that if you take audio tracks of dolphin conversations and compare them as graphic patterns of phonemes they strongly resemble our conversations.

One recent study that sought to test whether dolphins were speaking with one another in something like human language offered bottlenose dolphins a cache of treats. To obtain the delicacies the dolphins had to work together to open a container. Researchers found that in doing so the dolphins chattered frequently, using burst pulse signals, seemingly directing one another and jointly celebrating when the container was opened.

Herzing's own work turned up evidence that dolphins could be taught a made-up word that they adopted into their own vocabulary. The word was "sargassum." It's a type of seaweed that the dolphins like to play with. Herzing and her assistants joined in the dolphins' games of grasping and swimming about with the sargassum. Soon the dolphin researchers picked up on their hydrophones a dolphin reference to the sargassum in what appeared to be a slightly higher-pitched version of the word she had provided them. Whether mockingly or earnestly, the dolphins used this when they went back to playing with the clump of seaweed. While we do not yet know where their equivalent parts to the Broca's and Wernicke's areas—the language regions of our brain—are found, it matters little where they are. Language by its nature requires a different type of mind with more awareness for semantic cues, and a greater awareness for semantic cues necessarily implies more consciousness of the other individuals in the group.

There are other proofs of our likeness to cetaceans. Most notably, dolphins are the only other mammal species in which females always pass through menopause. It appears that they evolved for this for the same reasons we have. Dolphin grandmothers help raise the young, and it assists their chances for survival if the grandmother devotes herself to this rather than bearing more children. This will only occur with a creature that has a long life, one that includes an extended period of intensive rearing.[8]

In addition, like humans, dolphins are domesticated speakers who are tribal, poetic, and potentially suicidal, both as individuals and as members of groups. It has recently been shown that dolphins will kill their young. This is not filial cannibalism. Most often it happens when dolphin males kill dolphin babies unrelated to them in order to

---

8   Some primatologists have claimed that chimp females living in the wild also pass through menopause, and it has been observed in some captive chimps. However, as wild chimps are so dangerous and therefore so difficult to examine and observe, nothing can be definitively said about this.

get a female to ovulate. This is something that has been observed in many species, most famously among lions. But there is another kind of dolphin killing that is similar to a self-sacrificing human impulse. In captivity a dolphin female may kill her own child, seemingly to save it from a life without freedom.

Obviously, dolphins are very different from humans. That is shown in the organization and wiring of their brains. Where ours have been fashioned with an emphasis upon sight, theirs focus upon hearing. This is because they are so highly dependent upon their skills at echo-location. Moreover, dolphins can turn off one side of their brain and go to sleep, leaving the other half of the brain on. They are also almost without a sense of smell, and the structure of their brains reflects that. As they have just a vestigial olfactory bulb and it falls away soon after birth, there's nothing in the way of a nerve structure in their minds for processing scent.

They are also unlike humans in important respects with regard to behavior. Children are reared by groups of females, and then mature males leave and form into bachelor pods. I do not wish to overstate this comparison. I do not mean to suggest that humans are just like dolphins. Nor am I saying that we are herd creatures just as much as deer or sheep. My suggestion is rather that we need to extend our view of humans. We have long known that as *Homo* evolved our cerebral and prefrontal cortexes multiplied in size and changed in structure. This happened for a reason.

Much is made of man's creation of the tools for fire. Yet, within a tribe, fire is a far less dangerous instrument than words. Use of this tool prompts much more mayhem, far more conflict. Inherent in that—set in our brains alongside it—is a measure of social awareness far more sophisticated than that of our primate ancestors. Thus, language is not only a destructive and divisive instrument but also a tool for cooperation. Yet to employ it in this fashion we must be skilled at listening and following directions as well as talking. These are primate attributes but not in the same measure. These are instead the signal qualities of domesticated animals. We know from research into the taming of silver foxes that the ability to read social cues is an essential inherited characteristic associated with the process of domestication.

How did we become tame? Is it not unreasonable to conclude that it is a requirement of a species that is possessed of the power of speech? As we developed language, we had to evolve to make use of our capacity for it and for its growing complexity, and with this we began to take on some of the characteristics of the other large-brained mammals on the earth, all of which are herd creatures. We became more docile and less violent and also less independent in our nature and our judgments. Because of our terrestrial existence and our patterns of obtaining food, most *Homo sapiens* do not trek about as a herd. At least this is no longer the manner of our movements. It was, however, the manner in which traditional hunter-gatherer peoples went from place to place, migrating according to the seasons and the movements of the game they hunted. This is when most human evolution occurred. A similar process took place among cetaceans as they evolved.

As we are human, we are intensely social and passionately attached to the groups that provide us with our identity. Our thinking reflects that—for better and for worse. This draws us to groupthink and ideology over and above the search for truth or reliance upon logic. In the Middle East today, the greatest divide focuses upon the matter of who ought to have been selected as caliph 1,350 years ago. Battles are fought over this, and people are killed based on where they stand on this ancient question. Attachment to causes—and to faiths of all kinds—is strongly connected to the process of speech and language.

We can see that in the briefest examination of the foundational texts of the world's major religions. The first of the Ten Commandments is "I am the Lord thy God, thou shalt not have any gods before Me." The second is "Thou shalt not take the name of the Lord thy God in vain." Likewise, in Islam, the first of the Five Pillars is a statement of faith: "There is no God but Allah, and Mohammed is his prophet." Zoroastrianism's motto adheres to a similar logic: "Right thoughts, right words, right deeds." In each, concepts of speech and language precede calls to action, and the dictates of power follow not from evolutionary goals but from personal expressions: words.

I first presented my idea that the advent of spoken language has driven our evolution in the direction of greater and greater sociability in a novel I have been working on entitled *Herd Immunity*. I referred to it as the Childers-Khan hypothesis, naming it after one of the book's characters. The heroine is a cetacean researcher, and she comes through her work to believe—as I do—that in critical ways human psychology bears a greater correspondence to that of dolphins than it does to that of chimps and gorillas. In arguing this, I was following others. British academic Andrew Whiten may be the most prominent of the neuroscientists to make this argument. Another is Esther Nimchinsky. A professor at Rutgers University in New Jersey, she has often published her research in journals of primatology. Yet in a famous and much-cited 2007 paper entitled "Cetaceans Have Complex Brains for Complex Cognition," Nimchinsky joined with a group of leading cetacean specialists to acknowledge the fact that it makes at least as much sense to study dolphins and whales for the purpose of understanding human intelligence as apes.

One might argue in response that the capacity for spoken language and the sociability and social acuity it requires is no guarantee of our empathy and altruism. To that end, one could point to high-functioning sociopaths who are articulate and who display relatively good or even superior social skills. One could thereby postulate a path of human evolution in which most people had the lack of feeling for other people that chimps and gorillas display for one another but in which spoken language developed as a central feature of human society.

There are several problems with this idea. First, while sociopaths can possess these traits, a society mostly composed of them would not be cooperative. It would be one in which there was almost no trust, and it would be difficult if not impossible for its members to execute the tasks needed for human survival. As we have pointed out, our species faced special problems, and this required a predisposition for collaboration. Gangs of criminals tend to turn on each other, they split up regularly, and they depend upon theft and exploitation rather than production and creation. If everyone were like a mob boss, we would have all died off long ago. Remember: On several occasions, we barely survived with an opposite nature. Absent of some amount of green beard effect, we would have failed as hunter-gatherers. We would have been as helpless as a rat cornered by a fox. Think of homeless people. They can and do starve because of their isolation and their difficulty working together, and their survival is dependent upon support by others, whether obtained through begging or from philanthropic outreach. A *Homo sapiens* species made up of such atomized individuals would have been obliterated. Moreover, while it is undeniably *possible* to be articulate and sociopathic—as confidence men sometimes are—the growth of all those special features in the brain, like mirror neurons, assists in our understanding of language, *and* it makes it easier for us to grasp others' thoughts and feelings. Returning to our example of those suffering from autism, we know that these are connected. In fact, they are literally so in that they attach common neural pathways. These make us kinder and more gregarious. But they do not make us freer, rather the opposite.

This logic applies to the other species that depend upon some measure of spoken language. Like us, they are naturally skilled at cooperation. Without exception, they are also herd animals. The contrast with primates is striking. I have previously commented that orangutans are in some measure more like humans than chimps as they are more docile. Yet orangutans are not highly cooperative, and once they have reached maturity they rarely employ their voices to further cooperation with other adults of their own sex. As a text on primates and their behavior pointedly notes, "A fifteen year old male orangutan can make himself heard at a distance of one to two miles...." This is not to encourage mutuality. It is used to "provoke any young males to flee."

By and large, oral expression among apes is for conveying threats and fending off competition. Our anatomical likeness to primates conceals a series of fundamental dissimilarities. If we are to think of other animals as sisters and brothers, we would do better to look to the creatures with trunks and flippers.

Our impulse to speak arose at the same time as our inclinations to make art and music. That connects us to another human impulse: religiosity. We know that because so much of the outstanding art and music produced by man was stirred by religious faith. That lit the fire to the construction of our magnificent churches, temples, and mosques, and were we to make a roster of the greatest poets, composers, and painters we could not but be struck by how many were devout. It inspired the ancient Greek dramas, Dante's *Divine Comedy*, Bach and Mozart's masses and chorales, and Michelangelo's, Leonardo's, and Rembrandt's depictions of the Bible stories.

When Percy Bysshe Shelley referred to poets as "the unacknowl-edged legislators of the world," he was striking at something unde-niable. Great poets and artists aim to provide their audience with a message, one that serves a collective purpose. Through their art they aspire to explain abiding questions about the meaning of life. In this respect, the purpose of art is inherently cooperative, even if it is the expression of an individual. The great dramas of ancient Greece and Elizabethan England were written to bring together a city or a nation, retelling basic myths and a shared history.

This appears to be one of the earliest functions of art. Excavations of ancient cave dwellings show lapidary works that express a common purpose and meaning. Depicting herds of desirable game animals, they unite people in a moment of achievement or speak of collec-tive aims. In other cases, we find primitive sculptures that represent beloved deities or that serve as votive figures and funerary objects. That dolphins not only sing but teach one another songs that have meaning for their clan shows that these impulses are not unique to humans.

We have already mentioned that humans are much more inter-ested in intention and motivation than chimps. This hints at another aim of art. Poems, songs, and plays are a form of storytelling. Indeed, even instrumental music tells a kind of tale, and traditional forms like ballads perform this function. We can see this in popular tunes (e.g., "My Name Is Luka," "The Pina Colada Song"). This further demonstrates our need not only for distraction and amusement but also our yearning for connection.

CHAPTER 13

CAUSES OF SUICIDE

IT MAY BE IMPOSSIBLE TO understand human nature to any meaningful degree without considering the problem of suicide. A peculiarly human dilemma, it's frighteningly common. As violent as some American cities are, at least two and a half times as many Americans kill themselves every year as are murdered. Moreover, in other nations the record is worse. Japan does not even crack the top ten countries in suicide rate, but it is the leading cause of death among Japanese men aged twenty to forty-four. Indeed, more than twenty times as many Japanese perish each year by suicide as homicide, and suicide rates are even higher in South Korea, Russia, and India.

Yet suicide is not observed only among humans. It is an affliction of whales and dolphins. Examples of mass self-destruction among cetaceans turned up in Iran in 2007 and in Chile in 2015. In the first case, 152 dolphins beached themselves. In the second, 337 lifeless whales were found on the shore of a Patagonian fjord. Approximately 2,000 cetaceans beach themselves annually. Some researchers have attributed this to the effects of sonar, but events of this kind date back centuries before it was invented. During the 1500s there were reports of whales in the Netherlands stranding themselves. By the 1930s knowledge of this was sufficiently commonplace that newspaper reporters were filing stories about why they took place.

Some animal researchers suggest that this may not be collective suicide. But accepting that belief requires us to assume that exceptionally intelligent animals that can precisely plot out journeys of thousands of miles do not know what they are doing when they strand themselves.

This said, sea lions also engage in collective self-destruction, and there is evidence that those that beach themselves are suffering from the effects of toxic algae. This may cause damage to the hippocampus. That affects their mental state and sense of direction. It can even induce epileptic seizures. In rare cases stranded dolphins may be suffering from the same syndrome. Yet this doesn't seem to be what's happening most of the time when dolphins and whales beach themselves, and that stranded sea lions are tormented by environmentally-induced epilepsy isn't proof that they aren't acting deliberately and purposefully to end their lives. Nor does it explain the long-standing phenomenon of whales and dolphins engaging in communal acts of self-destruction.

The first writer to make suicide his chief area of study was the French sociologist Émile Durkheim. Durkheim was dismissive of the idea that animals could engage in self-destruction. Still, his book *Suicide* remains the foundational examination of the subject, and his concern with the social dimensions of the act is critical to gaining insight into its causes.

Durkheim divided suicides into four types. We have already mentioned altruistic suicide. Durkheim understood that this occurs most frequently in societies with intense social bonds and a strong concept of shame. Thus, it was seen in Imperial Japan as it had earlier occurred in ancient Rome. In the liberated societies of the contemporary West, Durkheim said that there was a much more common kind of suicide, which he termed *egoistic*. In this, the person ending his life is acting selfishly through an act of will. Though this might be prompted by an intense feeling of sadness or depression, at its heart is a person's diminished sense of connection and attachment to the world. Since men tend to be more egotistical and less socially engaged, it follows that they commit the greater number of these suicides. Though women may threaten to kill themselves, men carry it through. That imbalance is magnified as men tend to pick more effective means: all too often guns, rather than pills. Between three-fourths and four-fifths of suicides are of males.

Durkheim drew his conclusions from demographic statistics, but one of his most important claims was unsupported by data. This was that women's suicide rates should correlate inversely with the number of children they had. That is so. One study in Taiwan found that women with one child had a rate of eleven suicides per one hundred thousand per year, women with two children had seven, and women with three or more children had fewer than six. One might argue, of course, that it is a sign of mental health and overall contentment that a woman has additional children and that this reflects correlation rather than causation. Most researchers have concluded, though, that the sense of duty and purpose provided by children significantly reduces the risk of suicide.

There may be physiological reasons for this. Prolactin is a third important sex hormone. Since its primary function is to regulate lactation, women have higher levels than men. Abnormally high prolactin levels induce depression. In general, however, lactation seems to be beneficial for mental health as prolactin regulates and suppresses the hypothalamus, pituitary, and adrenal glands: the so-called HPA axis. Without the influence of prolactin, these boost cortisol levels, increasing feelings of stress. In effect, it seems that prolactin steadies women as they deal with the highs, lows, fears, and frustrations of newborns.

Obviously, Durkheim did not know that, and he worked from official statistics. But those told him much else. In particular, he was able to see that Protestant regions tended to have higher rates of suicide than Catholic areas. From this, Durkheim deduced that religious attitudes and beliefs could reduce its frequency and that the influence of example had a potent effect as well. That irreligion increases rates of suicide is shown by the high rates of suicide in highly secular countries like China and Russia, and in the markedly elevated rates of suicide among young people in the United States raised in secular homes. Conversely, religious faith and the demands of work appear to be deterrents against the feelings of meaninglessness that lead to suicide. That was indicated by a study on suicide in forty-two different countries. It found that people living in religious societies with bonds of connection among the faithful were less likely to kill themselves. The absence of these may partly explain another type of suicide that Durkheim described. He called this type *anomic*, and he thought of it as a response to a depression that arises among those who have a keen sense they have been cast out and rejected.

Durkheim did not investigate the times of year when people kill themselves, but there is research on this. Suicides rates rise in the spring and early summer. The general belief is that people kill themselves not merely because they are miserable, but when they see happy people around them. The sight of young couples holding hands is more apt to push a wavering figure over the edge than to inspire gladness.

Mass suicides are inspired by the feelings that awaken the impulse toward altruistic suicide and another kind of suicide that Durkheim referred to as *fatalistic*. These suicides occur among people who believe that the world is oppressing them and they have no future within it. All these facts direct us toward a larger point: Suicide is the act of a creature whose sense of himself is largely, though not exclusively, arrived at by how he thinks others see him—or fail to see him.

The most famous book on the subject is Leo Tolstoy's *Anna Karenina*. As Tolstoy may have been the most psychological of all novelists, it may be worth considering. Much remembered are its opening lines: "All happy families are alike. All unhappy families are different in their own way." Tolstoy began the novel in this way as one of his aims in writing it was to underscore the value of an age-old proposition. This is that a person's satisfaction and sense of purpose in life is most often connected to stable, supportive, affectionate bonds of matrimony.

The book contrasts two couples. Its secondary characters, Kitty and Levin, are well-suited to one another and tolerant of each other's faults. Tolstoy's beautiful but not always sensible heroine, Anna, is unhappily wed to a much older, unprepossessing councillor of state, Alexander Karenin. Meeting the handsome, cynical army officer, Alexander Vronsky, Anna finds herself drawn to him, and, imagining Vronsky to be a more soulful figure than he actually is, she commences an affair. The scandal precipitated by her infidelity, which she makes little effort to hide, causes her to lose the custody of her child and to be thrown out of high society. Things go from bad to worse as she tries to assuage her suffering by ingesting small amounts of morphine. Rapidly becoming addicted and no longer at peace with Vronsky, Anna throws herself in front of a train and kills herself.

Tolstoy depicted the characters he knew best: attractive figures from the Russian gentry, the aristocracy, and the military. Nonetheless, the difficulties he was writing about apply to every social class and country.

Both Tolstoy and Durkheim provide us with a trail of clues about how to prevent suicides. While Tolstoy was writing *Anna Karenina*, his interest in religion grew, as did his belief in the value of piety. He connected the success of Levin's marriage to Kitty with Levin finding faith, and he saw how Anna's suicide was related to her isolation and drug use. Anna's suicide is the egoistic act of a woman who is vain, conscious of her great beauty and addicted to drugs. It is anomic in that she has become a pariah, has been deprived of her child, and is living far away from her old life in Moscow and Saint Petersburg. It is fatalistic as it is a response to a sudden change in Vronsky's manners, one that makes her fearful that worse lies ahead in their affair. It is equal parts tragedy and case study, and Tolstoy depicts the psychology of a victim—a human victim—with faultless precision.

It does not answer the question, though, of why whales and dolphins kill themselves. But there are clues here, as well. One relates to those species most likely to strand themselves. Baleen whales, it turns out, are less inclined to engage in strandings than other whales. This makes sense as they are not cooperative hunters and need not be so in tune with those around them. It's also worth pointing out that there are credible reports of dogs committing suicide when their owners left them for extended periods. That's suggestive as dogs are also cooperative hunters—as we were for most of our evolutionary history.

What does biochemistry tell us? Dolphins have similar if slightly lower levels of testosterone than humans. One study found that two mature bottlenose males had approximately 10 percent lower levels of testosterone in their blood than an average human male. However, dolphins are more like humans—and less like chimpanzees—in that there are very dramatic differences in the ratio of testosterone levels between males and females.

Both sexes of dolphins are more docile than we are. Thus, we can teach them to behave like seals, performing tricks for food. That we may watch this and feel shame, seeing it as unworthy of them and unjust, shows that we are in some measure alike. This mutuality animates our common impulse toward self-destruction and our common search for identity.

A team of researchers who looked at nineteen dolphins that stranded themselves in Ireland in 1997 found that only one was ill. It seems that the rest followed the sick dolphin to shore. Reflecting on this, science writer Laurel Braitman observed:

> No other group of mammals has evolved in a space so devoid of spots to hide from predators. Dolphins and whales don't retreat to dens and burrows; they don't climb trees or hide in caves. In the face of danger, they are able to hide only behind one another. This may have affected their social worlds, making their ability to trust, communicate, and cooperate with one another even more important. This may also explain why some strandings include otherwise healthy individuals. These healthy dolphins and whales may strand simply because their social bonds with their ailing fellows are too powerful to allow them to swim away.

Humans appear to have more egotism than dolphins do, and we are not as tame. But the people who engage in mass suicide are likely acting as dolphins do when they collectively strand themselves. That would explain the folk who half-knowingly swallowed cups of cyanide-laced Kool-Aid as the members of the Jonestown Cult in Guyana did. Lacking in independence, they are those most in need of the support and the sense of identity provided by the group.

At the beginning of this book, I mentioned two more distinctive and strikingly peculiar behaviors by which people seek to destroy themselves or those around them: anorexia nervosa and Munchausen syndrome by proxy. Both reflect our attenuated sense of self. Because we are not as self-assured as apes are but we are far more social, we can be utterly desperate in our yearning for attention. In anorexia, these feelings act in concert, affecting young women who are unsure of their place. Those affected are more likely to suffer from obsessive-compulsive disorder (OCD). 41 percent of those diagnosed with anorexia or bulimia will also be diagnosed at some point in their life with OCD, and they are more than three times as likely to have a close relative who suffers from it. Frightened by the changes overtaking their bodies, they are eager to entice yet uncomfortable with their newfound sexuality and the power that potentially comes with it, and, under this fixation, they starve themselves and then gaze into a false mirror in which they see themselves as fashionably thin, almost sexless and deliriously elegant.

In Munchausen syndrome by proxy, the wish to be noticed matches itself to psychopathy. In seeking out sympathy and a perverse acclaim, mothers poison their own children, and nurses and doctors inject deadly toxins into their patients. Through this latter example, we see how a small number of humans are like chimps—showing their characteristic want of empathy—but utterly different, displaying a person's need for reaction from the group and for identity. These combine with a pernicious talent for planning.

CHAPTER 14

WHY PRIMATOLOGISTS MONKEY AROUND

THERE ARE MANY REASONS WHY primate anthropologists are so determined in their insistence that the essential nature of our species mirrors that of primates. One is the rough similarity in our skeletal structures. Another is self-interest, a force of which they are largely unconscious. There is also an uncritical assumption that this must be so given our shared, if distant, genetic lineage with chimps and gorillas. A fourth reason derives from their experience with baby and infant chimps.

You may have read that Michael Jackson had a baby chimp named Bubbles and that he liked to play with him. It is no secret that Jackson was lonely, and he appears to have developed a real bond with Bubbles. This is not unusual. If you look online, you can find a touching picture of Frans de Waal holding a baby chimp, and on the cover of the paperback edition of Jane Goodall's *In the Shadow of Man* there's a picture showing her reaching out her hand to an infant chimp. I think it's a safe bet that nearly all chimp researchers formed an attachment at some point with an immature chimp. These relationships spur feelings of commonality and empathy of the kind one inevitably will experience with a cat or a dog, and they can't help but distort their perceptions of them. This is in spite of the fact that these types of relationships cannot be sustained into the creature's adulthood when, as we know, they become exceedingly dangerous.[9]

---

9   Bubbles was eventually given up for adoption to a primate research center. This incoherence is reflected in reporting on chimps. In November 2021, the Wellington Zoo in New Zealand attracted attention across the South Seas nation with news of the birth of an "adorable" chimp on its premises. Yet a few years earlier, zoo officials had revealed to the press that chimps were first on their list of animals kept on a "shoot to kill list" should they ever escape.

Several more causes come to mind, which we have not touched on. One relates to facial expression. A region of the brain that is most alike in humans and apes is the part of the prefrontal cortex that controls it. Chimps have very similar muscles controlling facial expression as humans, and though they do not show the extreme tendency toward right-handedness that humans do, they are alike to us in that their facial expressions appear more on one side of the face than the other. This evolutionary inheritance has come to us mostly unimpaired, and it makes us suppose that primate thoughts are as much like ours as primate facial expressions. (This fallacy applies within our species as well. We may think that a sociopath is thinking as we are because he smiles, winks, and grimaces in the same way.) Yet research on chimp facial expressions has shown that the repertoire of expressions that they typically display is different from ours. As you might expect, they are far more likely to show their bared teeth or to scream and hoot. What's more, when chimps and other primates make eye contact, it's usually interpreted not as communication but as a threat. This has been found to be especially true of chimps and gorillas with the result that they tend to avoid sustained eye contact with one another.

Nonetheless, since their facial *structures* are so similar to ours, they have the same basic "vocabulary" of facial expressions available to them. They can frown, grin, and nod at each other just as we do. So even though they are most likely to use this repertoire to convey their instincts toward dominance and aggression and we are most apt to use them to assist in collaboration and to facilitate intimacy, they appear alike. From this outward similarity, we can be drawn toward a mistaken faith in an inmost kinship.

A fifth reason is related to our intelligence. Since cleverness is the most salient human attribute, we are apt to assume that it must be the thing to focus on in trying to understand our evolution. But, as we have seen, human evolution wasn't spurred by this gift. It was thrust ahead by our weaknesses. It was our unsuitability for our new role as hunters that forced the change in our nature and led to our higher

236

intelligence. In effect, primate researchers are reversing matters. They assume that as chimps have moderately large brains that we must be extending an existing direction and pattern in evolution when in actuality we are parting from a prior evolutionary line.

Another reason may be less than obvious. As recently as 50,000 years ago, there were at least four different species of *Homo*. Not only were *Homo sapiens* roaming the earth, but there were still Neanderthals and Denisovans. There was even a dwarf humanoid species in Indonesia known as *Homo floresiensis*, sometimes called Hobbit Man. And if you go back a quarter million years, there were probably a dozen different hominin species. Were these all still among us it's likely that anthropologists would more readily accept the idea that we should be placed in our own distinct order. That we—*Homo sapiens*—alone survived pushes those in the business of classifying animals toward the notion that we should be placed within an existing order, even though we are no longer much like them and lack for a defining characteristic of the order: prehensile feet, used for climbing trees and grasping branches. This impulse to think of humans as primates was defensible when less was known about primate behavior. It is an ironic but undeniable fact that those who have done the most to promote the notion of our kinship have played an essential part in disproving the claim. Their field research has shown how murderous and self-interested chimps are, how they do not commit suicide or exhibit preferential homosexual instincts, how much they sleep and what they consume, how little they cooperate, how they struggle to listen or follow directions, how much slower their metabolism is, and many of their other fundamental dissimilarities to us.

There is even one more reason to consider for the unwillingness of so many anthropologists to acknowledge the enormous gulf between humans and apes. We may think that primates are like us because of the ape sense of humor. It has to be pointed out here that when we use the term "humor" we actually mean two things, though we generally do not bother distinguishing between them. One type of humor is that displayed by a class clown. This is the wit of the person making the joke, and it tends to carry an undercurrent of aggression. The second meaning is responsiveness to a jest. In some respects, this is the opposite.

A clue that is so is found in the reactions of professional comics to witticisms. It's proverbial that while they make it their business to tell jokes that they almost never laugh at them. My friend Roger Simon offers a proof. A successful Hollywood screenwriter of the 1970s and 1980s, Simon wound up writing a movie for comedian Richard Pryor. Most critics and audiences say that Pryor was the greatest of all stand-up comics. Yet my friend reports that when he and Pryor met each week to go over the script they were working on, Pryor didn't just refuse to laugh. Simon claims that he wouldn't even shift the muscles in his face to acknowledge jokes. Judged by responsiveness, the funniest man in the world had no sense for comedy. Most of us display that second kind of humor when we laugh at our boss's not particularly clever remarks. This is the behavior of a socially adept creature able to respond in a nonthreatening way.

While no other animal vocalizes laughter quite as humans do, most mammalian species have some form of it. In some cases, they shake. In dogs, it's shown by a breathy pant that every dog owner recognizes. In dolphins, it's a squeal followed by a whistle. When rats are tickled, they give off a high-pitched squeak.

It's not clear that chimps are any more inclined to playful laughter of this type than dogs or dolphins are. But, when they do laugh, it seems both visually and audibly more like our chuckling. More persuasively, they are fond of the first type of humor. They like tomfoolery. If given a seltzer bottle, they will spray it on people. Provided with a stick, they may throw it back at us. This minor character trait has provided primatologists with the final piece of evidence persuading them of what I call the primate myth: the idea that we can learn who we are by studying chimps and gorillas and that our underlying patterns of behavior are the same.

A striking example of how profoundly the ape sense of humor and the similarity in his facial expressions influences our perception of them is given in the book *Reaching the Animal Mind*. Its author, Karen Pryor (no relation to Richard), was among the most influential experts on animal training of the last century. Her particular interests were dogs and dolphins. Since chimpanzees are neither tame nor good at taking instruction from humans, the book mentions them only in passing on a single page. She does, however, include a brief vignette in which she describes her experiences with a gorilla kept at Disney's Animal Kingdom in Orlando. Struck by his impishness and his impulse to make a "funny face," she concludes her account of meeting him by saying that "I am used to thinking of animals as individuals, which, of course, they are; but this is a little different. A *person* is inside that huge creature: a primitive one, but a person. Furthermore, he's a comedian."

This statement was ironic as Pryor's work was intended to show our commonality with other creatures through their ability to follow instructions: something chimps and gorillas are comparatively poor at. Moreover, if an animal's impulse to engage in playful teasing was really proof that it was much alike to us, then we would be very much like capuchin monkeys and parrots, both of which display this instinct too. In any event, no ape makes sophisticated jokes of the kind we do.

CHAPTER 15

MAIN LESSONS OF EVOLUTION—INCLUDING HOW AND WHY WE HAVE MISUNDERSTOOD THE CAUSES OF WAR AND PEACE

PHILOSOPHERS FREQUENTLY DISTINGUISH BETWEEN "NATURAL man" and "civilized man." This conception of the problem goes back to Hobbes and Rousseau. The dichotomy suggests that civilization has made us obedient and tame but that we are not "naturally" so in its absence and in some essential way that we are just like the apes who routinely kill one another. The difficulty with this idea is screamingly apparent: How are humans capable of civilized existence if it is not within our nature? Plainly, the development of advanced societies is something we are capable of that apes are not. To the same degree, our propensity for suicide and warfare are distinctively human problems. I do not mean to suggest that humans won't be especially cruel in the absence of civil order. Plainly, in a condition of anarchy people will act barbarically. However, that does not mean that we are "naturally" merciless or vicious. We are not chimps.

With that in mind, let's consider the statement that the humans most like chimps are psychopaths. This is an apt description of most dictators and warlords. Take, as an example, Napoleon Bonaparte. His brutality was almost routine. At his command thousands of enemy prisoners were "run through": bayoneted to death. He also ordered the gassing of multitudes of rebellious Black slaves held in French ships, and he called for French military doctors to give fatal overdoses of opiates to his soldiers in Syria who could not be evacuated. What distinguished him from most would-be conquerors was not any goodness of heart but his intelligence and his dexterity as a politician and a general.

The interest and fascination with Napoleon extends beyond this to such much talked-about subjects as his height and his two marriages. Yet these are not the most important facts about the French ruler. The key point is that he was a dictator—autocrats like to wage war, and, as we are a herd animal, we are inclined to follow them. So, if we give our power over to despots, wars will result. We can see this through Napoleon and how the French acted when he was their leader.

French attitudes toward Napoleon changed after his loss at the Battle of Leipzig in 1814. Sometimes called the Battle of the Nations, Leipzig was a calamitous French defeat that led to the restoration of the Bourbon monarchy. With its return, Napoleon was placed in exile on the isle of Elba. The new government, headed by Louis XVIII, was seen by many in France as a puppet regime, attached as he was to the British king. So there continued to be support throughout France for Napoleon. Yet the sympathies of the people were divided. The overwhelming majority of the men wished for Napoleon's return. Women favored Louis XVIII.

But as most people are natural conformists, hardly anyone would openly speak out for Napoleon while Louis XVIII was in power, and when word reached Paris in March 1815 that Napoleon had escaped from Elba and was returning, the first newspaper headlines depicted this as a terrifying threat. Over the following days, though, the tenor of the stories changed, and by the time he arrived in the capital nearly every influential man was claiming that he had been among the emperor's most dogged advocates.

Back in command, Napoleon raised an army and set out on a new plan of conquest: the so-called 100 Days campaign. This concluded with his greatest loss, delivered at the Battle of Waterloo. That was followed by his second exile, in which he was imprisoned on a small island in the South Atlantic as another restoration of the Bourbon king, Louis XVIII, took place.

Napoleon's wars led to the death or disability of more than one-fourth of the draft-age male population of France. Most historians maintain that this remarkable toll demolished the fighting spirit of France. The evidence suggests otherwise. As long as Napoleon was in charge, young men were inspired. They saw this as an instrument for advancement. Wars meant the opportunity for honors and titles. They permitted low-ranking peasants to meet and mate with foreign women. They offered the opportunity for plunder. They were what a generation of French men was taught to fantasize about. French women saw things differently. To them war was the reason they had lost brothers and could not find husbands. Can we find a helpful contrast to this? One is close at hand. Our country's history is illuminating.

Many books that have the word "democracy" in their title fail to distinguish between the different types of it. The United States has had several. In its initial form, the Constitution did not call for a democracy but rather a representative republic. The president was selected by designated figures of the establishment. The chosen electors were not obligated to vote for specific candidates, and United States senators were selected by state legislators.

Matters soon changed though. The Twelfth Amendment, passed in 1804, made the presidency an office picked by a broad mass of the citizenry. This was Jeffersonian democracy. In the Northern states, property-owning males could vote. In the South, only property-owning White men could. Whether either form of government was really democracy is a valid question. What is certain is that the North and South had different attitudes and beliefs regarding war. Both during the War of 1812 and the Mexican-American War, the South was enthusiastic about it. The North was not. Indeed, during the War of 1812, opposed to the fighting as many Northerners were, Northern leaders convened to discuss secession.

Another set of reforms was enacted in the aftermath of the Civil War. All men were given the right to vote by the Fifteenth Amendment. Then, during the Spanish-American War, the country became inflamed by reports of Spanish atrocities in Cuba, and the jingoistic accounts of the American newspapers convinced their readers that Spain had plotted an attack against the United States that injured our honor. The subsequent conflict was a curious one. The United States pointedly refused to take possession of Cuba, the seeming jewel we had conquered, and, though we committed many atrocities in the Philippines, we started the islands out on a deliberate process to prepare them for self-rule. With slavery eliminated, the voters still considered fighting to be glorious, but the grosser dimensions of it—rule over others and open theft of their lands—were seen as shameful.

A truer democracy appeared with the passage of the Nineteenth Amendment, women's suffrage. That followed the Seventeenth Amendment, which arranged for direct election of senators. Combining the two, the United States became a democracy in which the voters pick the most important representatives and in which the electorate is composed of law-abiding adults of both sexes. As we have already noted, no democracy of this kind has ever made war against another. So, the problem of armed struggle between nations becomes a problem not merely of human nature but of the manner of our governments, and we must conclude that this form of democracy is critically important to the peaceful survival of our species.

But, if we cannot make every country into a democracy in which women hold a voice, can we employ trade and diplomacy to prevent armed struggles?

In the last century, two colossal international organizations were set up to deter war: the League of Nations and the United Nations. If you assume that war reflects the human impulses of primates but that wise men can solve these problems, then their failure must seem like a great mystery. If, however, you take for granted that many of the diplomats serving in these institutions were not enlightened, rational figures but merely the servile henchmen of despotic criminals, then there is no riddle.

One source for this delusion was the American Secretary of State Henry Kissinger. As a historian he promoted what might be called the Metternich Myth. This is the notion that the Austrian Count Klemens von Metternich had constructed a durable system of alliances that prevented a recurrence of the large-scale fighting of the Napoleonic Wars. Supposedly, Metternich's brilliant arrangement of affairs at the Congress of Vienna set Europe up for a long period of peace. In fact, that assembly of European monarchs didn't even sustain a peace while it was in session. Napoleon's return and the Battle of Waterloo took place as it was happening. Moreover, while it is true that during the next hundred years there was little direct conflict among the European powers, Metternich's diplomacy had little to do with this. Rather, the Europeans states were busy with wars against native tribes in their newly created colonies in Africa and Asia, while Britain, France, and Germany were undergoing a process of democratization. So, their fighting with one another was limited to relatively brief conflicts like the Crimean War and the Franco-Prussian War. But, once belligerent rulers assumed power in Germany, two world wars followed.

Similarly, Kissinger's own diplomacy at the Paris Peace Conferences did not conclude a war. Although it was the window dressing for an American withdrawal from Vietnam, it happened even as Ho Chi Minh's tank battalions were conquering the south. Minh's armies then invaded Laos and Cambodia. A few years later they fought a brief but bloody war with China. The Nobel Peace Prize given to Kissinger produced no peace. Similarly, the diplomats who served Hitler and Stalin didn't stop any wars. Nor could they have. Just the opposite: They assisted in plotting the Second World War by writing the Molotov-Ribbentrop Pact, the agreement by which Nazi Germany and the Soviet Union allied themselves. This is typical. A diplomat employed by a psychopath isn't likely to gently coax him out of his elaborately conceived plans of attack, and an institution like the UN composed of men representing such rulers can't prevent them. "Wise men" inflate their accomplishments and their capacities. Wars will only end when evil men no longer have the power to start them. By creating democracies in which women have a voice, we have stopped wars and set aside those capable of manipulating us into them. That is the path to ending war. The primate model of our species has led us to false ideas about the extent to which we can trust undemocratic foreign powers. We are seeing that once more in Ukraine as we may soon witness it with respect to China's desire to retake Taiwan.

Trade is not an answer to this problem either, and it may even worsen the problem. The First World War took place in an era of increasing trade. This not only didn't prevent fighting, it may have hastened it. Japan's attack on Pearl Harbor was a direct consequence of the decision of a diplomat, American Secretary of State Cordell Hull, to cut off its access to US oil supplies. In the same way, the rising tension with China has run parallel to increased commerce. To stop wars we must first acknowledge their real cause: Our tameness makes us susceptible to the war fever promoted by autocrats, as it makes use of young men ready and willing to follow their commands.

The primate model of our species inclines us toward the belief that humans will always be crooked, rational, and individualistic. Yet few people are consistently any of these things. The mistaken belief that we are not a herd animal also causes us to underestimate the risks of panic selling and buying in financial markets, and it leads us to absurd notions about marriage, fidelity, and the importance of stable social bonds. Altogether, it serves as a partial explanation for the failure of our elites to think substantively about the risks, problems, and issues that face us.

Humanity will benefit when we relinquish the notion that we are apes wearing collared shirts. Our species has derived little advantage from it, and while the example of many crime gangs and a few corporate boardrooms may comport with the theory, far too many human behaviors conflict with it for it to be terribly helpful. In addition, as it encourages people to think of human relations as inherently transactional, it makes them more so. Rather than be seen as primates, we should be placed in a separate order of mammals: *Homo*. At a bare minimum, we should be in a distinct family grouping.

We can see how ridiculous the present view of our species is by looking at the classification of koalas and kangaroos. Koalas are marsupials. They have much smaller brains than kangaroos, they live in trees, and their diet is almost exclusively composed of eucalyptus leaves. Like most arboreal mammals, they sleep most of the day. Kangaroos are also marsupials, but they are terrestrial, they form into herds called mobs, they sleep less than half as much as koalas, and they have much larger brains and a far more diverse diet. Consequently, the two groups are broken up into separate families: Phascolarctidae (koalas) and Macropodidae (kangaroos and wallabies).

Wholesale reclassification has previously occurred within the primate order. For almost half a century, the order had six families within it, rather than the present five. That was because it included tree shrews. These creatures, native to South America, are closely related to lemurs, and they have prehensile feet. However, as they are adorned with claws instead of nails, they have recently been reclassified within a separate order called Scandentia.

In the same way, humans fail to meet the definitional requirements of the primate order as we do not have prehensile feet. Yet we are classified within the family Hominidae: the great apes. This is though the differences between humans, and chimps, gorillas, and orangutans are far greater as only humans are meso-carnivores, only humans are tame, only humans speak to one another, and only humans are neonates. To the same degree, only *Homo sapiens* engage in religious sacrifices; follow fashion trends; create financial bubbles and panics; have strong linguistic identification and impulses toward linguistic nationalism; start wars; become fascinated by sports teams; are inclined toward preferential homosexuality; impose ostracism and slavery on others; and venerate priestly celibacy, monogamous mating, and dozens of other uniquely human patterns of behavior. These start to make sense if we reject the false and simplistic view of humans that we have been asked to adopt: that we are close kin of chimps.

We are not the savage animals they are. The average person spends more time worrying about what his neighbors think about his lawn ornaments than he does contemplating their murder. Moreover, under the right circumstances the number of people occupying themselves with such bloodthirsty schemes could be far fewer. That would require not only more effective policing but also more two-parent families and better schools. In any event, as we have noted, suicide ends more lives than homicide, and wars and governments kill millions more.

But, in one way, it makes sense that we are classed in the same order and family as the great apes: It's another proof of how much illogic is a natural human attribute and how innately drawn we are toward received ideas and to a credulous regard for popular suppositions.

One person who grasped this aspect of our nature was the philosopher Bertrand Russell. Realizing that our thought was only occasionally guided by logic, Russell suggested that schools and colleges ought to spend more time teaching it. Ironically, he offered this counsel during the precise era in which advocates of progressive education were removing logic—the study of fallacies and syllogisms—from the school curriculum. This deprived students of a formal course in the subject, something that had been part of their education since the Middle Ages. Nonetheless, as Russell himself acknowledged, more study of logical argument and reasoning would only marginally improve the quality of our thought. That's because our impulse to herd thinking is so strong.

I opened this book by examining the life and work of the man who commenced the investigation of the human genome: Theophilus Painter. In Painter we see much of what is best in mankind and some of the worst. By most accounts, he was a dutiful husband, a good father, and an encouraging teacher. Enlisting in the Army during the First World War, he served in the Reserve and departed it with the rank of captain. A talented gardener, he crafted a number of novel varieties of irises. A devoted member of his church congregation, he was one of the outstanding scientists of the twentieth century. Yet in his experiments he made use of materials taken through involuntary castrations, and as a university administrator he upheld segregation. That support for inequality continued for the length of time when the idea was broadly ratified among the class in which he lived. In Painter, we see much that is admirable, along with our enormous capacity for blindness and indifference to the problems and concerns of those we believe are unlike us. Even so, because Painter lived in a democracy, there was a limit to the amount of wrongdoing he could engage in. Had he been raised in a society like Nazi Germany or the Soviet Union, the more unsavory aspects of his nature would have been put to use for far greater evil.

We are more than a half century past Painter's death, and Richard Dawkins is now the best-known living theoretician of the human genome. Dawkins's fame arose from his 1976 book *The Selfish Gene.* Within its pages, Dawkins put forward his belief in the paradox of altruism. This is the notion that altruism, if it exists at all, is a form of selfishness. As he put it, "pure, disinterested altruism, [is] something that has no place in nature, something that has never existed before in the whole history of the world." This is consistent with the book's larger thesis: Individual genes act for their own continuance, and they are independent of the organism of which they are a part. This is a controversial idea. One reason is the evidence that has appeared in recent decades for a biological basis for altruism among humans and other animals. Undoubtedly, Dawkins would respond to this by arguing that this is selfishness in disguise. His view is that we are altruistic because this advances genes for altruism, not because there is such a thing as a separate concept of goodness. However, in putting forward this belief, Dawkins is making a philosophical argument that obscures practical reality. He becomes a little like the person who says that the extinction of the sun as it transforms into a white dwarf negates the value of the light outside our window. While the statement may prove true in a few billion years, it is not of any use as a guide to existence and more revealing about the nature of the speaker than the world in which we live. It is a reduction to absurdity.

Elsewhere in the book, Dawkins introduced the term "meme." This is the concept of an aspect of culture that can be transmitted. This is ironic. For among the most influential memes of recent decades is the idea that humans are apes, and Dawkins is both a fervent adherent of this notion and a determined believer that human problems can be solved exclusively through our powers of reason. Dawkins has also become famous for his opposition to religion, which he sees as hostile to reason, and for his advocacy for atheism. Yet Dawkins's arguments against belief in god are not founded in definitive evidence. They are a form of faith. Moreover, his certainty in his own powers of reasoning are belied by his persistent commitment to the idea that mankind can best be understood as a species of ape. This meme has taken over his brain and prevented him from properly interpreting the subject whose investigation has been the work of his entire life. Even so, he believes that he can successfully employ the same mechanism of analysis to answer questions that would appear to be beyond human reasoning.

His errors lead us toward one conclusion: There is something to be said for maintaining and fostering an amount of traditional belief. After all, if people are naturally inclined to follow and obey and are as easily deluded as an exceptionally clever man like Dawkins has been in the precise area of his supposed expertise, then keeping some established rules and principles of behavior and devotion is desirable. The alternative path—radical reconstruction of society and of our ideals—is dangerous.

Dawkins has frequently argued that society must "prevent" religious parents from "indoctrinating" their children in their faith. To achieve this end, he would have governments given more authority. But the combined force of disappearance of traditional beliefs and the growth of governmental power has all too often led to totalitarianism, and the autocrats guiding these states have shown themselves to be as fond of war as all dictators are. This is no recipe for a better world. As much as we are inclined toward obedience, almost any of us might be drawn into the service of a malign cause or government were we within its herd.

No matter what we conclude about these questions, though, it is worth noting that all of them are irrelevant to chimps and gorillas. They are pertinent to us because our evolutionary past has drawn us far away from them. Acknowledging that is a first step to understanding ourselves. Those who believe that we are very much like primates serve as a peculiar proof that we are different from them as their thinking has been taken, without meaningful reflection, from the herd.

ENDNOTES AND SOURCES

CHAPTER 1: "THE PRIMATE MYTH"

Painter's new knife: Bentley Glass, "Theophilus Shickel Painter: 1889–1969," *Biographical Memoir*, National Academy of Sciences, Washington, DC, 1990, p. 314.

Painter's study of an opossum: Dorothy Haskett, "Theophilus Shickel Painter (1889–1969)," *The Embryo Project Encyclopedia,* https://embryo.asu.edu/items/172873, https://embryo.asu.edu/pages/theophilus-shickel-painter-1889-1969.

Painter's source for his human chromosome slides: Wilson John Wall, *The Search for Human Chromosomes: A History of Discovery*, Spencer International Publishing, Switzerland, 2015, pp. 64–66. See also Glass, "Theophilus Shickel Painter: 1889–1969," p. 315.

Painter's conclusion of twenty-four human chromosome pairs: Valerie J. Grant, *Maternal Personality: Evolution and the Sex Ratio—Do Mothers Control the Sex of the Infant*, Routledge Press, UK, 1998, p. 81.

Sweatt's work as a mailman: Gary M. Lavergne, *Before Brown: Heman Marion Sweatt, Thurgood Marshall, and the Long Road to Justice*, University of Texas Press, Austin, Texas, 2010, pp. 17–18.

Sweatt's work as a newspaper columnist: Max Krochmal, *Blue Texas: The Making of a Multi-Racial Democratic Coalition in the Civil Rights Era*, University of North Carolina Press, Chapel Hill, 2016, p. 70.

Painter's presentation of data on black and white chromosome pairs: Dan Graur, "Theophilus Painter: The Geneticist on the Wrong Side of History," *Judge Starling Blog*, July 8, 2020, https://judgestarling.tumblr.com/post/623095087703965696/theophilus-painter-the-geneticist-on-the-wrong.

Painter's belief in segregation: Amilcar Shabazz, *Advancing Democracy: African Americans and the Struggle for Access and Equity in Higher Education in Texas*, University of North Carolina Press, Chapel Hill, 2004, pp. 66–94.

Primates have prehensile feet: John P. Rafferty, ed., *Primates: The Britannica Guide to Predators and Prey*, Britannica Educational Publishing, 2011, New York, p. 24. A common misconception is that all primates have fully prehensile hands. In fact, many tarsiers and marmosets do not. The characteristic and uniform attribute of the order is prehensile *feet*. These are necessary for climbing trees. As such, Wikipedia recently defined a primate by noting that "all [primate] species have adaptations for climbing trees." Humans, of course, do not have an adaptation for climbing trees.

Claim that chimps are more like humans than gorillas: Jane Goodall, *In the Shadow of Man*, First Mariner Books, 2000, Boston, pp. 251 and 282.

Primetime TV show about a family raising a chimp: Garry Marshall and Thomas L. Miller, *Me and the Chimp*, CBS, 1972, https://en.wikipedia.org/wiki/Me_and_the_Chimp.

Spanish parliamentary resolution on chimps: Lee Glendenning, "Spanish parliament approves 'human rights' for apes," *The Guardian*, June 26, 2008, https://www.theguardian.com/world/2008/jun/26/humanrights.animalwelfare.

EU legislation on apes: "EU bans the use of great apes in research," Project R&R, September 21, 2010, https://releasechimps.org/resources/article/eu-bans-great-apes-in-research.

Chimps alleged to be "nicer" than humans: Jane Goodall, *Through a Window: My Thirty Years with the Chimpanzees of Gombe*, Mariner Books, Houghton Mifflin Harcourt, Boston-New York, 2000, p. 123. She writes, "For so many years I had believed that chimpanzees, while showing uncanny similarities to humans in many ways, were by and large, rather 'nicer' than us." After more than a decade of studying chimps, she was surprised to discover that they had a "darker side to them."

Border collie skill in understanding human speech: John W. Pilley and Alliston K. Reid, "Border collie comprehends object names as verbal referents," *Behavioural Processes*, February 2011, vol. 86, no. 2, pp. 184–195. One set of researchers at Central Washington University claimed that the chimp known as Washoe was able to learn 350 words. But others who looked into this, including the famous investigator Thomas Sebeok, have suggested that the researchers were guilty of self-deception. Close examination of videotape of Washoe and other chimps instructed in sign language show that they were just copying their instructors or, at best, using a method to obtain an outcome but with no grasp of the actual meaning of the "words" used.

How chimps learn human language: Clive D. L. Wynne, *Do Animals Think?*, Princeton University Press, Princeton, NJ, 2006, p. 125.

Canine skill at learning human language: Christa Lesté-Lasserre, "Gifted dogs can learn 12 words in a week and remember them for months," *New Scientist*, October 6, 2021.

Chimp struggles with pattern recognition tests: Johan Lind, Magnus Enquist, and Stefano Ghirlanda, "Animal memory: A review of delayed matching-to-sample data," August 1, 2015, *Behavioural Processes*, vol. 117, pp. 52–58.

Assorted animals and the cooperative pulling test: James Gorman, "Wolves Know How to Work Together," *The New York Times*, November 7, 2017. See also Martin Schmelz, Shona Duguid, Manuel Bohn, and Christoph Völter, "University of Veterinary Medicine, Vienna Cooperative problem solving in giant otters (*Pteronura brasiliensis*) and Asian small-clawed otters (*Aonyx cinerea*)," *Animal Cognition*, November 2017, vol. 20., no. 1.

Chimp diet and proportion of meat consumed: James McWilliams, "This Absolute Lust for Meat," *Pacific Standard*, December 4, 2017. The author is quoting statistics provided by veteran chimp researcher Craig Stanford.

Bonobo and antelope meat consumption compared: Vicky M. Oelze, Benjamin T. Fuller, Michael P. Richards, Barbara Fruth, Martin

Surbeck, Jean-Jacques Hublin, and Gottfried Hohmann, "Exploring the contribution and significance of animal protein in the diet of bonobos by stable isotope ratio analysis of hair," *Proceedings of the National Academy of Sciences*, May 31, 2011, https://www.pnas.org/content/early/2011/05/24/1018502108.abstract.

Frans de Waal's misleading claims about the chimp diet: Frans de Waal, *Mama's Last Hug: Animal Emotions and What They Tell Us About Ourselves*, W. W. Norton, New York, 2020, p. 159.

De Waal's attempts to manipulate his subordinate's findings: I found this out during a personal interview with Professor James K. Rilling at his offices in Atlanta on the Emory University campus. Rilling is head of Emory's Laboratory for Darwinian Neuroscience. Rilling told me that de Waal insisted that no one working for him at the Emory National Primate Research Laboratory should ever claim that there was a difference between the human brain and a chimp brain unless they could definitively prove that this was so and that it was basic and fundamental. As Rilling is loyal to his former boss, the professor defended this. Yet all of Rilling's research has been revealing these profound differences.

Jane Goodall's statements about chimp cooperation: Goodall, *In the Shadow of Man*, p. 196.

Lack of evidence of chimp coordination using communication: Pavel V. Voinov, Josep Call, Günther Knoblich, Marina Oshkina, and Matthias Allritz, "Chimpanzee Coordination and Potential Communication in a Two-touchscreen Turn-taking Game," *Nature*, February 25, 2000, https://www.nature.com/articles/s41598-020-60307-9. See also Ian C. Gilby and Richard C. Connor, "The Role of Intelligence in Group Hunting: Are Chimpanzees Different from Other Social Predators?" chapter 18 in *The Mind of the Chimpanzee: Ecological and Environmental Perspectives*, edited by Elizabeth V. Lonsdorf, Stephen R. Ross, and Tetsuro Matsuzawa, University of Chicago Press, Chicago, 2010.

Poorer outcomes observed among larger chimp hunting groups: Ian C. Gilby, Lynn E. Eberly, Lilian Pintea, and Anne E. Pusey,

"Ecological and social influences on the hunting behaviour of wild chimpanzees, *Pan troglodytes schweinfurthii*," *Animal Behaviour*, July 2006, vol. 72, no. 1, pp. 169–180.

Chimp capacity for altruism shown in rope experiment: Keith Jensen, Brian Hare, Josep Call, and Michael Tomasello, "What's in it for me? Self-regard precludes altruism and spite in chimpanzees," *Proceedings of the Royal Society B: Biological Sciences*, April 22, 2006, vol. 273, no. 1589, pp. 1013–1021.

Infant and toddler food sharing: David F. Bjorklund, *How Children Invented Humanity: The Role of Development in Human Evolution*, Oxford University Press, UK, 2021, p. 221. Bjorklund is referencing Jensen, Hare, Call, and Tomasello. See also Rodolfo Cortes Barragan, Rechele Brooks, and Andrew N. Meltzoff, "Altruistic food sharing behavior by human infants after a hunger manipulation," *Scientific Reports*, vol. 10, no. 1785, February 4, 2020.

Chimp cannibalism: Claire A. Kirchoff, *Life and Death in the Gombe Chimpanzees: Skeletal Analysis as an Insight into Life History*, Springer International Publishing, 2019, p. 86.

Chimp murder rates as observed by anthropologists: Michael Balter, "Why do chimps kill each other?" *Science* magazine, September 17, 2014, https://www.sciencemag.org/news/2014/09/why-do-chimps-kill-each-other.

Brutality of langur monkeys discussed publicly: Frans de Waal, *Our Inner Ape: A Leading Primatologist Explains Why We Are Who We Are*, Riverhead Books, New York, 2005, p. 107.

Theophilus Painter agrees to support desegregation: Ricky F. Dobbs, *Yellow Dogs and Republicans: Allen Shivers and Texas Two-Party Politics*, Texas A&M University Press, College Station, p. 58.

Chimp and Human DNA Vary By 13.5%, Not 1.4%: Dong Ahn Yoo, Arang Rhie, Prajna Hebbar, Francesca Antonacci, Glennis A. Logsdon, Steven J. Solar, Dmitry Antipov, et al., "Complete Sequencing of Human Genome," *Nature*, April 9, 2025, vol. 641, pp. 401–418.

Human and chimp promoter genes vary widely: Ralph Haygood, Olivier Fedrigo, Brian Hanson, Ken-Daigoro Yokoyama, and Gregory A. Wray, "Promoter regions of many neural- and nutrition-related genes have experienced positive selection during human evolution," *Nature Genetics*, August 12, 2007, vol. 39, pp. 1140–1144.

Importance of indels in human DNA as seen in the Zoonomia Project: Carl Zimmer, "What Cheetahs, Armadillos and Whales Revealed About Human DNA," *The New York Times*, April 27, 2023. See also Sacha Vignieri, *Science*, April 27, 2023, vol. 380, no. 6643, pp. 356–357, DOI: 10.1126/science.adi1599.

Rate of change in the genetic clock: Vincent M. Sarich and A. C. Wilson, "Rates of albumin evolution in primates," *Proceedings of the National Academy of Sciences*, July 1967, vol. 58, no. 1, pp. 142–148. See also Navin Elango, James W. Thomas, and Soojin V. Yi, "Variable molecular clocks in hominoids," *Proceedings of the National Academy of Sciences*, vol. 103, no. 5, January 23, 2006, pp. 1370–1375.

Taming of the silver fox by Soviet scientists: Krishna Ramanujan, "Silver fox study reveals genetic clues to social behavior," phys.org, September 24, 2018, https://phys.org/news/silver-fox-study-reveals-genetic-clues-to-social-behavior.html. See also Jason G. Goldman, "Man's Best New Friend? A Forgotten Russian experiment in fox domestication," *Scientific American*, Frontier Blog Network, September 6, 2010.

Chimps have five times our pulling strength: Frans de Waal, *Our Inner Ape*, p. 138. Other researchers have suggested that although the difference is very great that it isn't quite so dramatic. Some other estimates are more in the range of three times as great.

Chimp brain mostly grows in the womb: Smithsonian Institution, "Human Characteristics: Brains—Bigger Brains: Complex Brains for a Complex World," http://humanorigins.si.edu/human-characteristics/brains.

Bonobo promiscuity: William Tucker, *Marriage and Civilization: How Monogamy Made Us Human*, Regnery Publishing, Washington, DC, 2014, p. 35.

Human and ape sexual dimorphism compared: Hector Reynoso, "Sexual Body Size Dimorphism," Center for Academic Research & Training in Anthropogeny, https://carta.anthropogeny.org/moca/topics/sexual-body-size-dimorphism.

Human versus ape likelihood of bearing twins: John J. Ely, William I. Frels, Sue Howell, M. Kay Izard, Michale E. Keeling, and D. Rick Lee, "Twinning and Heteropaternity in Chimpanzees (*Pan troglodytes*)," *American Journal of Physical Anthropology*, May 2006, vol. 130, no. 1, pp. 96–102.

Pair-bonding in different mammal species: Connor T. Lambert, Anne C. Sabol, and Nancy G. Solomon, "Genetic Monogamy in Socially Monogamous Mammals Is Primarily Predicted by Multiple Life History Factors: A Meta-Analysis," *Frontiers In Ecology and Evolution*, September 19, 2018. See also Ker Than, "Wild Sex: Where Monogamy Is Rare," LiveScience.com, November 20, 2006, https://www.livescience.com/1135-wild-sex-monogamy-rare.html.

Anorexia nervosa observed in ancient times: Pamela Norris, *Eve: A Biography*, NYU Press, New York, 2001, p. 212.

Parrots name their children: Karl S. Berg, Soraya Delgado, Kathryn A. Cortopassi, Steven R. Beissinger, and Jack W. Bradbury, "Vertical transmission of learned signatures in a wild parrot," *Proceedings of the Royal Society: Biological Sciences*, July 13, 2011, https://doi.org/10.1098/rspb.2011.0932.

Elephants appear to use names in communicating with one another: Michael A. Pardo, Kurt Fristrup, David S. Lolchuragi, Joyce H. Poole, Peter Granli, Cynthia Moss, Iain Douglas-Hamilton, and George Wittemyer, "African elephants address one another with individually specific name-like calls," *Nature: Ecology & Evolution*, published online June 10, 2024.

Chimp skill at remembering numbers: Sana Inoue and Tetsuro Matsuzawa, "Working memory of numerals in chimpanzees," *Current Biology*, vol. 17, no. 23, p. R1004, 2007, Primate Research Institute, Kyoto, Japan, https://www.cell.com/current-biology/pdf/S0960-9822(07)02088-X.pdf.

Frans de Waal's claims about the chimp capacity for showing empathy: De Waal, *Our Inner Ape*, p. 30.

Frans de Waal cites Thomas Hobbes to explain behavior of chimps and humans: De Waal, *Our Inner Ape*, p. 54.

Frans de Waal claims that bonobos are "gentle" creatures: De Waal, *Mama's Last Hug: Animal Emotions and What They Tell Us About Ourselves*, p. 197.

Frans de Waal claims that bonobos are "peaceful hippies": Henry Mance, "Frans de Waal: 'We are very much like primates,'" *The Financial Times*, March 8, 2019.

The Associated Press parrots de Waal's claims about bonobos: Seth Borenstein, "The bonobo, the non-murderous version of the chimpanzee, gets its genome mapped," *Christian Science Monitor*, June 13, 2012, https://www.csmonitor.com/Science/2012/0613/The-bonobo-the-non-murderous-version-of-the-chimpanzee-gets-its-genome-mapped.

Bonobos are capable of co-feeding: Elvio Angeloni, *Physical Anthropology: Annual Edition 07/08*, McGraw-Hill/Dushkin, Dubuque, Iowa, 2006, p. 55.

Bonobos are better at reading social cues but worse at making tools than Pan troglodyte *chimps:* Brian Hare and Richard Wrangham, "Equal, Similar but Different: Convergent Bonobos and Conserved Chimpanzees," in *Chimpanzees and Human Evolution*, edited by Martin N. Muller, Richard W. Wrangham, and David Pilbeam, Harvard University Press, Cambridge, Massachusetts, 2016, pp. 152–153.

Male-male fighting is more common among bonobos than Pan troglodyte *chimps:* Maud Mouginot, Michael L. Wilson, Nisarg Desai, and Martin Surbeck, "Differences in expression of male aggression between wild bonobos and chimpanzees," *Current Biology*, April 12, 2024.

Bonobos wake up ready to fight: Carl Zimmer, "No 'Hippie Ape': Bonobos Are Often Aggressive, Study Finds," *The New York Times*, April 12, 2024. The article is quoting Maud Mouginot, author of the previously cited comparative *Current Biology* field study.

Bonobos are one hundred times as violent as humans: Wrangham, pp. 19–20.

Best practices recommended by the European Zoo Association for handling bonobos and cheetahs: Jeroen Stevens, ed., "EAZA Best Practices: Bonobo (*Pan paniscus*)," European Association of Zoos and Aquaria, pp. 83–93, January 2020. See also Kimberly Sengenberger, Hilke Bus, and Lars Versteege (Southern Cheetah EEP coordinator), eds., "EAZA Best Practice Guidelines Cheetah (*Acinonyx jubatus*)," European Association of Zoos and Aquaria, p. 17, January 2020.

Neurological similarity of chimps and bonobos: William D. Hopkins, Heidi Lyn, and Claudio Cantalupo, "Volumetric and Lateralized Differences in Selected Brain Regions of Chimpanzees (*Pan troglodytes*) and Bonobos (*Pan paniscus*)," *American Journal of Primatology*, vol. 71, no. 12, December 2009, pp. 988–997.

Chimp sexual behavior during estrus: Christopher Mogielnicki and Katherine Pearl, "Hominid sexual nature," *Theory in Biosciences*, March 2020, vol. 139, no. 2, pp. 191–207.

Comparison of relative genetic closeness of chimps and bonobos to humans: Hare and Wrangham, p. 147.

Bonobos must be taught by humans how to hunt for ants: Hare and Wrangham, p. 146.

Bonobo filial cannibalism documented: Ewen Callaway, "Hippy apes caught cannibalising their young," *New Scientist*, February 1, 2010, https://www.newscientist.com/article/dn18454-hippy-apes-caught-cannibalising-their-young/.

Carl Linnaeus's claims about humans and primates: Justin E. H. Smith, *Nature, Human Nature, and Human Difference: Race In Early Modern Philosophy*, Princeton University Press, Princeton, New Jersey, 2016, p. 136. See also Adriana S. Benzaquén, *Encounters with Wild Children: Temptation and Disappointment in the Study of Human Nature*, McGill-Queen's University Press, Quebec, Canada, 2006, p. 125.

Darwin and suicide: Charles Darwin, *On the Origin of Species by Means of Natural Selection*, Grant Richards, Leicester Square, London, 1902, p. 181.

CHAPTER 2: "HOW EVOLUTION CHANGED THE HUMAN BRAIN"

EMI developed the first CT scanners: Bruce Hillman and Jeff Goldsmith, *The Sorcerer's Apprentice: How Medical Imaging Is Changing Health Care*, Oxford University Press, UK, 2010, p. 34. This claim has been questioned: See Zeev V. Maizlin and Patrick M. Vos, "Do we really need to thank the Beatles for the financing of the development of the computed tomography scanner?" *Journal of Computed Assisted Tomography*, March–April 2012, vol. 36, no. 2, pp. 161–164.

Few MRI scanners in 1993: Ellie Kincaid, "Want Fries with That? A Brief History of Medical MRI, Starting with a McDonald's," April 16, 2018, *Forbes*, https://www.forbes.com/sites/elliekincaid/2018/04/16/want-fries-with-that-a-brief-history-of-medical-mri-starting-with-a-mcdonalds/?sh=11cf907e3de0.

Frans de Waal's attacks on Steven Pinker: Mance, "Frans de Waal: 'We are very much like primates,'" *Financial Times*, March 8, 2019.

Steven Pinker explains fMRI's potential to Time: Steven Pinker, "Will the Mind Figure Out How the Brain Works?" *Time*, April 10, 2000, vol. 155, no. 14.

FMRI reproduces a Pink Floyd song: Philippe Albouy, Lucas Benjamin, Benjamin Morillon, and Robert J. Zatore, "Distinct sensitivity to spectrotemporal modulation supports brain asymmetry for speech and melody," *Science*, February 28, 2020, vol. 367, no. 6481, pp. 1043–1047. See also Hana Kiro, "Scientists Recreate Pink Floyd Song by Reading Brain Signals of Listeners," *The New York Times*, August 15, 2023.

Effect of brain size among primates: R. O. Deaner, K. Isler, J. Burkart, and C. van Schaik, "Overall brain size, and not encephalization quotient, best predicts cognitive ability across non-human primates," *Journal of Brain, Behavior and Evolution*, 2007, vol. 70, no. 2, pp. 115–124. See also Lori Marino, "Absolute brain size: Did we throw the baby out with the bathwater?," *Proceedings of the National Academy of Sciences*, October 2006, vol. 103, no. 37, pp. 13563–4.

Similarity of chimp and human brains with respect to sensory and motor functions: Todd M. Preuss, "The human brain: rewired and running hot," *Annals of the New York Academy of Sciences*, May 2011, vol. 1225 (supplement 1), pp. E182–E191.

Chimp brains are not asymmetrical in the manner of human brains: Li Xiang, Timothy Crow, and Neil Roberts, "Automatic analysis of cross-sectional cerebral asymmetry on 3D in vivo MRI scans of human and chimpanzee," *Journal of Neuroscience Research*, e-published February, 27, 2019, https://pubmed.ncbi.nlm.nih.gov/30809847/.

Human Broca's area is greatly expanded: Natalie M. Schenker, William D. Hopkins, Muhammad A. Spocter, Amy R. Garrison, Cheryl D. Stimpson, Joseph M. Erwin, Patrick R. Hof, and Chet C. Sherwood, "Broca's Area Homologue in Chimpanzees (*Pan troglodytes*): Probabilistic Mapping, Asymmetry, and Comparison to Humans," *Cerebral Cortex*, March 2010, vol. 20, no. 3, pp. 730–742, https://www.ncbi.nlm.nih.gov/pmc/articles/PMC2820707/.

Expansion of the arcuate fasciculus in humans: James K. Rilling, "Comparative primate neurobiology and the evolution of brain language systems," *Current Opinion in Neurobiology*, vol. 28, October 2014, p. 13.

Changes in the human inferior frontal gyrus: Dirk Jan Ardesch, Lianne H. Scholtens, Longchuan Li, Todd M. Preuss, James K. Rilling, and Martijn P. van den Heuvel, "Evolutionary expansion of connectivity between multimodal association areas in the human brain compared with chimpanzees," *Proceedings of the National Academy of Sciences*, April 2, 2019, vol. 116, no. 14, pp. 7101–7106.

Loss of empathy shown when the supramarginal gyrus is shut off: Christopher Bergland, "The Neuroscience of Empathy," *Psychology Today*, October 10, 2013, https://www.psychologytoday.com/us/blog/the-athletes-way/201310/the-neuroscience-empathy.

Connection shown between a poorly functioning cingulate gyrus and Tourette's syndrome and ADHD: John M. Allman, Atiya Hakee, Joseph M. Erwin, Esther Nimchinsky, and Patrick Hof, "The Anterior Cingulate Cortex: The Evolution of an Interface Between Emotion and Cognition," Annals of the New York Academy of Sciences, May 2001, vol. 935, pp. 114.

Relationship between a poorly functioning cingulate gyrus and criminality: Nobuhito Abe, Joshua D. Greene, and Kent A. Kiehl, "Reduced engagement of the anterior cingulate cortex in the dishonest decision-making of incarcerated psychopaths," *Social Cognitive and Affective Neuroscience*, September 5, 2018, https://pubmed.ncbi.nlm.nih.gov/29982639/.

Relationship between a weak cingulate gyrus and recidivism: Eyal Aharoni, Gina M. Vincent, Carla L. Harenski, Vince D. Calhoun, Walter Sinnott-Armstrong, Michael S. Gazzaniga, and Kent A. Kiehl, "Neuroprediction of Future Rearrest," *Proceedings of the National Academy of Sciences*, April 9, 2013, https://www.pnas.org/content/110/15/6223.

Relationship between the size of the anterior cingulate gyrus and empathy: Isabella Mutschler, Céline Reinbold, Johanna Wankerl, Erich Seifritz, and Tonio Ball, "Structural basis of empathy and the domain general region in the anterior insular cortex," *Frontiers in Human Neuroscience*, May 8, 2013, vol. 7, https://www.frontiersin.org/articles/10.3389/fnhum.2013.00177/full.

Differences in size of the anterior cingulate gyrus between men and women: Jing Liu, Jon-Kar Zubieta, and Mary Heitzeg, "Sex differences in anterior cingulate cortex activation during impulse inhibition and behavioral correlates," *Psychiatry Research*, January 30, 2012, vol. 201, no. 1, pp. 54–62.

Experiment showing the effect of a muted supramarginal gyrus on rat behavior: Julen Hernandez-Lallement, Augustine Triumph Attah, Efe Soyman, Cindy M. Pinhal, Valeria Gazzola, and Christian Keysers, "Harm to Others Acts as a Negative Reinforcer in Rats," *Current Biology*, March 23, 2020, vol. 30, no. 6, pp. 949–961, https://www.cell.com/current-biology/fulltext/S0960-9822(20)30017-8#%20.

Impulse control and the functioning of the cingulate gyrus: Luqing Wei, Nana Guo, Chris Baeken, Minghua Bi, Xiaowan Wang, Jiang Qiu, and Guo-Rong Wu, "Grey Matter Volumes in the Executive Attention System Predict Individual Differences in Effortful Control in Young Adults," *Brain Topography*, January 2019, https://pubmed.ncbi.nlm.nih.gov/30203260/.

Gray matter in the cingulate gyrus and the likelihood of mental disorders: Jing Lou, Yueji Sun, Zhixi Cui, and Lei Gong, "Common and distinct patterns of gray matter alterations in borderline personality disorder and posttraumatic stress disorder: A dual meta-analysis," *Neuroscience Letters*, January 10, 2021, vol. 741. See also Malin Höistad, Devorah Segal, Nagahide Takahashi, Takeshi Sakurai, Joseph D. Buxbaum, and Patrick R. Hof, "Linking white and grey matter in schizophrenia: oligodendrocyte and neuron pathology in the prefrontal cortex," *Frontiers in Neuroanatomy*, July 2, 2009, vol. 3, https://doi.org/10.3389/neuro.05.009.2009.

High levels of gray matter in the cingulate gyrus of chimps and indifference to others: William D. Hopkins and Jared P. Taglialatela, "Initiation of Joint Attention Is Associated with Morphometric Variation in the Anterior Cingulate Cortex of Chimpanzees (*Pan troglodytes*)," *American Journal of Primatology*, January 8, 2013. The authors believe that the high levels of gray matter in chimp anterior cingulate gyrus are reducing their ability to work with and respond to others. They suggest that this should also be true in humans, although, as noted above, the opposite appears to be the case: High levels of gray matter in the anterior cingulate of humans are associated with healthy interpersonal skills and overall mental health.

Frans de Waal's attacks on discussion and consideration of the relevance of the cingulate gyrus to human-chimp differences: De Waal, *Mama's Last Hug,* p. 222.

Relationship between psychopathy and poor function in the amygdala: Julian C. Motzkin, Joseph P. Newman, Kent A. Kiehl, and Michael Koenigs, "Reduced Prefrontal Connectivity in Psychopathy," *The Journal of Neuroscience,* November 30, 2011, vol. 31, no. 48, pp. 17348–17357.

Relationship between psychopathy and overall size of the amygdala: Abigail A. Marsh, Elizabeth C. Finger, Derek G. V. Mitchell, Marguerite Reid Schneider, Courtney Sims, David Kosson, Kenneth E. Towbin, Ellen Leibenluft, Daniel S. Pine, and Robert James R. Blair, "Reduced Amygdala Response to Fearful Expressions in Children and Adolescents with Callous-Unemotional Traits and Disruptive Behavior Disorders," *American Journal of Psychiatry,* July 2008, vol. 165, no. 6, pp. 712–720. See also Yaling Yang, Adrian Raine, Katherine L. Narr, Patrick Colletti, and Dr. Arthur W. Toga, "Localization of Deformations within the Amygdala in Individuals with Psychopathy," *Archives of General Psychiatry,* September 2009, vol. 66, no. 9, pp. 986–994.

Proposed connection between the development of the amygdala and human evolution: Caroline H. Lew, Kari L. Hanson, Kimberly M. Groeniger, Demi Greiner, Deion Cuevas, Branka Hrvoj-Mihic, Cynthia M. Schumann, and Katerina Semendeferi, "Serotonergic innervation of the human amygdala and evolutionary implications," *American Journal of Physical Anthropology,* November 2019, vol. 170, no. 3, pp. 351–360.

Development of the striatum and thalamus upon human evolution: Mary Ann Raghanti, Melissa K. Edler, Alexa R. Stephenson, Lakaléa J. Wilson, William D. Hopkins, John J. Ely, Joseph M. Erwin, Bob Jacobs, Patrick R. Hof, and Chet C. Sherwood, "Human-specific increase of dopaminergic innervation in a striatal region associated with speech and language: a comparative analysis of the primate basal ganglia," *Journal of Comparative Neurology*, July 1, 2016, vol. 524, no. 10, pp. 2117–2129. See also Mary Ann Raghanti, Melissa K. Edler, Alexa R. Stephenson, Emily L. Munger, Bob Jacobs, Patrick R. Hof, Chet C. Sherwood, Ralph L. Holloway, and C. Owen Lovejoy, "A neurochemical hypothesis for the origin of hominids," *Proceedings of the National Academy of Sciences*, January 2018.

Effect of spoken language upon the canine caudate nucleus: Ashley Prichard, Peter F. Cook, Mark Spivak, Raveena Chhibber, and Gregory S. Berns, "Awake fMRI Reveals Brain Regions for Novel Word Detection in Dogs," *Frontiers in Neuroscience*, October 15, 2018, https://www.frontiersin.org/articles/10.3389/fnins.2018.00737/full.

Function of the dorsal medial prefrontal cortex and capacity to work with strangers in chimps and humans: Sarah K. Barks, Lisa A. Parr, and James K. Rilling, "The Default Mode Network in Chimpanzees (*Pan troglodytes*) Is Similar to That of Humans," *Cerebral Cortex*, September 2013, vol. 3, no. 2, pp. 538–544.

Action of Brodmann's area 10 in apes and humans: Katerina Semendeferi, Este Armstrong, Axel Schleicher, Karl Zilles, and Gary W. Van Hoesen, "A Comparative Study of Area 10," *American Journal of Physical Anthropology*, vol. 114, pp. 224–241, 2001.

Size and role of the prefrontal cortex in apes, humans, and cetaceans compared: Alejandro Chinea, "Is Cetacean Intelligence Special? New Perspectives on the Debate," *Entropy*, October 13, 2017, vol. 19, no. 10, *Special Issue: Entropy and Its Applications Across Disciplines*, https://www.mdpi.com/1099-4300/19/10/543.

Gene pair changes in HAR1: Katherine S. Pollard, Sofie R. Salama, Bryan King, Andrew D. Kern, Tim Dreszer, Sol Katzman, Adam Siepel, Jakob S. Pedersen, Gill Bejerano, Robert Baertsch, Kate R. Rosenbloom, Jim Kent, and David Haussler, "Forces Shaping the Fastest Evolving Regions in the Human Genome," *PLOS Genetics,* October 13, 2006, https://doi.org/10.1371/journal.pgen.0020168.

Effect of calretinin on the cingulate gyrus: Patrick R. Hof, Esther A. Nimchinsky, Daniel P. Perl, and Joseph M. Erwin, "An unusual population of pyramidal neurons in the anterior cingulate cortex of hominids contains the calcium-binding protein calretinin," *Neuroscience Letters,* vol. 307 (2001), p. 141.

Changes in the LOXL2 gene and their implications for human and ape brain development: GeneCards: The Human Gene Database: LOXL2, LOXL2 Gene - GeneCards | LOXL2 Protein | LOXL2 Antibody.

Mirror neurons and the theory of mind: Antonella Tramacere, Telmo Pievani, and Pier Francesco Ferrari, "Mirror neurons in the tree of life: mosaic evolution, plasticity and exaptation of sensorimotor matching responses," *Biological Reviews of the Cambridge Philosophical Society,* November 16, 2016, vol. 92, no. 3, pp. 1819–1841.

Relative discovery of mirror neurons in chimps and humans: A. O. Ibegbu, U. E. Umana, W. O. Hamman, and A. S. Adamu, "Von Economo neurons: A Review of the Anatomy and Functions," *Austin Journal of Anatomy,* November 10, 2014, https://austinpublishinggroup.com/anatomy/fulltext/Anatomy-v1-id1026.php.

Numbers of von Economo neurons found in humans and apes: John M. Allman, Karli K. Watson, Nicole A. Tetreault, and Atiya Y. Hakeem, "Intuition and autism: a possible role for Von Economo neurons," *Trends In Cognitive Sciences,* August 2005, vol. 9, no. 8, pp. 367–373.

Von Economo neurons in humans and whales: Andy Coghlan, "Whales boast the brain cells that 'make us human,'" *New Scientist,* November 27, 2006, https://www.newscientist.com/

article/dn10661-whales-boast-the-brain-cells-that-make-us-human/#ixzz7GM88nsYZ. See also J. F. Prather, S. Peters, S. Nowicki, and R. Mooney, "Precise auditory–vocal mirroring in neurons for learned vocal communication," *Nature*, January 17, 2008, vol. 451, no. 7176, pp. 305–310.

Relationship between lack of von Economo neurons and autism spectrum disorders in humans: Michael R. McGowen, Lawrence I. Grossman, and Derek E. Wildman, "Dolphin genome provides evidence for adaptive evolution of nervous system genes and a molecular rate slowdown," *Proceedings of the Royal Society B: Biological Sciences*, July 18, 2012, https://doi.org/10.1098/rspb.2012.0869.

CHAPTER 3: "WHY HUMANS FIGHT WARS"

Population of chimps in the wild: James Eng, "Captive Chimps Join Their Wild Cousins in Getting 'Endangered' Status: Captive chimpanzees deserve the same protections as wild chimps, federal wildlife officials say," NBC News, June 12, 2015.

Use of marine mammals in warfare: Jane J. Lee, "Military whales and dolphins: What do they do and who uses them?" *National Geographic*, May 3, 2019, https://www.nationalgeographic.com/animals/article/140328-navy-dolphin-sea-lion-combat-ocean-animal-science?loggedin=true. See also Verity Murphy, "Let Slip the Sea Lions of War," BBC News, March 11, 2003. See also H. I. Sutton, "Trained Russian Navy Dolphins are Protecting Black Sea Naval Base, Satellite Photos Show," USNI News, April 27, 2022, https://news.usni.org/2022/04/27/trained-russian-navy-dolphins-are-protecting-black-sea-naval-base-satellite-photos-show.

Killer whales known to engage in coordinated attacks on human targets: Emma Ogao, "Killer whales learn 'coordinated' attacks on sailboats, some observers say," ABC News, May 23, 2024.

Ant wars described: E. O. Wilson, *Tales from the Ant World*, Liveright Publishing, New York, 2020, p. 2.

Facts about the Milgram experiments: Peter Lunt, *Stanley Milgram: Understanding Obedience and Its Implications*, Palgrave Macmillan, London, UK, 2009, p. 42.

Number of troops on the Russo-Chinese border: Benjamin Brimelow, "A bloody battle over a tiny island raised fears that China and the Soviets would start World War III," Business Insider, March 10, 2021.

History of Sino-Soviet conflict and how it almost led to nuclear war in 1969: Andrew Osborn and Peter Foster, "USSR Planned Nuclear Attack on China in 1969," *The Daily Telegraph*, May 13, 2020.

CHAPTER 4: "THE TWO GREAT WAVES OF HUMAN EVOLUTION"

Role of diet as foundational determinant of evolution: "How Darwin's Finches Got Their Beaks: A Gene's Eye View of Evolution," *Harvard Magazine*, July 2006.

Relationship between intelligence and meat-eating among animals: Johan Lind, Magnus Enquist, and Stefano Ghirlanda, "Animal memory: A review of delayed matching-to-sample data," *Behavioural Processes*, August 1, 2015, vol. 117, pp. 52–58.

Number of tigers in America as compared with those still in the wild: Faith Karimi, "There are more tigers in captivity in the US than in the wild," CNN, May 13, 2021, https://www.cnn.com/2021/05/13/us/tigers-captive-us-wild-trnd/index.html.

Human and ape propensity for making eye-contact: Fumihiro Kano and Josep Call, "Cross-species variation in gaze following and conspecific preference among great apes, human infants and adults," *Animal Behaviour*, May 2014, vol. 91, p. 143. See also Michael Tomasello, Brian Hare, Hagen Lehmann, and Josep Call, "Reliance on head versus eyes in the gaze following of great apes and human infants: the cooperative eye hypothesis," *Journal of Human Evolution*, March 2007, vol. 52, no. 3, pp. 314–20.

Chart of growth of brain and body size among prehistoric hominids: Russell Howard Tuttle, "Human Evolution: Increasing Brain Size," *Encyclopedia Britannica*, https://www.britannica.com/science/human-evolution/Refinements-in-tool-design.

Appearance of FOXP2 gene in the human genome: Michael Balter, "'Speech Gene' Debut Times to Modern Humans," *Science* magazine, August 14, 2002, https://www.sciencemag.org/news/2002/08/speech-gene-debut-timed-modern-humans.

Failure of speech among those without working FOXP2 genes: Michael C. Corballis, *The Truth About Language: What It Is and Where It Came From*, University of Chicago Press, Chicago, 2017, p. 166.

Mitochondrial DNA and presence of Neanderthal speech genes: Cosimo Posth, Christoph Wißing, Keiko Kitagawa, Luca Pagani, Laura van Holstein, Fernando Racimo, Kurt Wehrberger, Nicholas J. Conard, Claus Joachim Kind, Hervé Bocherens, and Johannes Krause, "Deeply divergent archaic mitochondrial genome provides lower time boundary for African gene flow into Neanderthals," *Nature Communications*, July 4, 2017, vol. 8, no. 16046.

Incest observed among Neanderthals: Australian Broadcasting Corporation, "Neanderthals' incest, interspecies sex revealed by genome," December 23, 2013, https://www.cbc.ca/news/science/neanderthals-incest-interspecies-sex-revealed-by-genome-1.2474245.

Uniformity of Neanderthal tool design: Jill Rubalcaba and Peter Robertshaw, *Every Bone Tells a Story: Hominin Discoveries, Deductions, and Debates*, Charlesbridge, Watertown, Massachusetts, 2010, pp. 79–80.

Changes in human dentition as evidence of self-domestication: William White Howells, *Getting Here: The Story of Human Evolution*, Compass Press/Howells House, Washington, DC, 1997, p. 237.

Human sexual dimorphism established long ago: John Noble Wilford, "New Clues to History of Male and Female," *The New York Times,* section C, p. 1, August 26, 1997, https://www.nytimes.com/1997/08/26/science/new-clues-to-history-of-male-and-female.html.

Percentage of caloric intake used by the brain: Marcus E. Raichle and Debra A. Gusnard, "Appraising the brain's energy budget," *Proceedings of the National Academy of Sciences,* August 6, 2002, vol. 99, no. 16, pp. 10237–10239.

Glucose distribution as a cause of apes' greater physical strength: Katarzyna Bozek, Yuning Wei, Zheng Yan, Xiling Liu, Jieyi Xiong, Masahiro Sugimoto, Masaru Tomita, Svante Pääbo, Raik Pieszek, Chet C. Sherwood, Patrick R. Hof, John J. Ely, Dirk Steinhauser, Lothar Willmitzer, Jens Bangsbo, Ola Hansson, Josep Call, Patrick Giavalisco, and Philipp Khaitovich, "Exceptional Evolutionary Divergence of Human Muscle and Brain Metabolomes Parallels Human Cognitive and Physical Uniqueness," *PLOS Biology,* vol. 12, no. 5, May 2014, https://pubmed.ncbi.nlm.nih.gov/24866127/.

Chimp propensity for biting both chimps and humans: Amy Hanes, "For Chimps, Human Touch Can Hurt," *Sapiens* magazine, November 2, 2018, https://www.sapiens.org/culture/chimpanzee-human-touch/. See also G. A. Engel and L. Jones-Engel, "Primates and primatologists: social contexts for interspecies pathogen transmission," *American Journal of Primatology,* June 2012, vol. 74, no. 6, pp. 543–50.

Geographical distribution of primates and reptiles: Pandey and Mathur, *Biology of Chordates,* PHI Learning, Delhi, India, 2018, p. 504.

Strength of carnivore jaws: Renee Robinette Ha, Tracy L. Brad, and James C. Ha, *Breed Differences in Dog Behavior: Why Tails Wag Differently,* CABI, Oxfordshire, UK, 2004, p. 195. See also David Macdonald, *The Velvet Claw: A Natural History of the Carnivores,* Parkwest, New York, 1992, p. 118. A spotted hyena's jaw can exert a pressure of 1,140 pounds per square inch. The human jaw can only produce pressure of roughly 160 pounds per square inch.

Descriptions of big cat predation upon chimps: Christophe Boesch, "The effects of leopard predation on grouping patterns in forest chimpanzees," *Behaviour*, 1991, vol. 117, no. 3–4, p. 226–230.

Past near extinction of Homo sapiens *as indicated by gene tests:* Robert Krulwich, "How Human Beings Almost Vanished from Earth in 70,000 B.C.," October 22, 2012, NPR.com. The article is quoting Sam Kean's *The Violinist's Thumb: And Other Lost Tales of Love, War and Genius as Written by Our Genetic Code*, Little Brown and Company, New York, 2012.

Desmond Morris's theories about human evolution discussed: Desmond Morris, *The Naked Ape*, Random House, New York, 2010, pp. 35, 45, and 124–160.

Chimp inability to engage in stone-knapping: Blandine Bril, Jeroen Smaers, James Steele, Robert Rein, Tetsushi Nonaka, Gilles Dietrich, Elena Biryukova, Satoshi Hirata, and Valentine Roux, "Functional mastery of percussive technology in nut-cracking and stone-flaking actions: experimental comparison and implications for the evolution of the human brain," *Philosophical Transactions of the Royal Society B: Biological Sciences*, January 12, 2012, https://doi.org/10.1098/rstb.2011.0147.

Differences in brain design among whale species: R. Douglas Fields, "Are Whales Smarter Than We Are?" *Scientific American*, January 15, 2008.

Range of elephant calls: Carl Safina, *Beyond Words: What Animals Think and Feel*, Henry Holt, New York, 2015, p. 92.

Biochemistry and its role in human and ape reactions to carrion: Caleb E. Finch, "Evolution of the human lifespan and diseases of aging: Roles of infection, inflammation, and nutrition," *Proceedings of the National Academy of Sciences*, October 12, 2009. See also Sonia Ragir, Martin Rosenberg, and Philip Tierno, "Gut Morphology and the Avoidance of Carrion Among Chimpanzees, Baboons, and Early Hominids," *Journal of Anthropological Research* (Winter 2000), vol. 56, no. 4, pp. 477–512.

ApoE lipoprotein presence in sea mammals: R. W. Davis, V. R. Pierotti, S. J. Lauer, S. T. Hubl, J. W. McLean, J. L. Witztum, and S. G. Young, "Lipoproteins in pinnipeds: analysis of a high molecular weight form of apolipoprotein E," *Journal of Lipid Research*, June 1991, vol. 32, no. 6, pp.1013–1023.

CHAPTER 5: "WHY WE ARE DIFFERENT FROM MOST CARNIVORES?"

Sleep patterns of lions: Mark C. Ross and David Reesor, *Predator: Life and Death in the African Bush*, Harry N. Abrams, New York, 2007, p. 146.

Sleep patterns of humans and primates compared: Jen Viegas, "How Human Sleep Differs from Other Primates, and What It Means for Our Health," Seeker.com, February 14, 2018. See also Jeffrey M. Jones, "In U.S., 40% Get Less Than Recommended Amount of Sleep," Gallup.com, December 19, 2013.

Sleep patterns of ungulates: "The Wildlife Society 2020: How Mammals Avoid Getting Caught Sleeping," Joshua Rapp Learn, November 5, 2020, https://wildlife.org/tws2020-how-mammals-avoid-getting-caught-sleeping. See https://scienceblogs.com/tetrapodzoology.com/2008/09/15/natural-history-of-sleep

Humans as meso-carnivores based on observed data of hunter-gatherer societies: Loren Cordain, Janette Brand Miller, S. Boyd Eaton, Neil Mann, Susanne H. A. Holt, and John D. Speth, "Plant-animal subsistence ratios and macronutrient energy estimations in worldwide hunter-gatherer diets," *The American Journal of Clinical Nutrition*, March 2000, vol. 71, no. 3, pp. 682–692.

Amount of meat-intake of primates: Steven W. Gangestad, "Reproductive Strategies and Tactics," *Oxford Handbook of Evolutionary Psychology*, edited by Robin Ian MacDonald Dunbar and Louise Barret, Oxford University Press, UK, 2007, p. 326.

Meat consumption levels of black bears: Linda Masterson, *Living with Bears: A Practical Guide to Bear Country*, PixyJack Books, Masonville, Michigan, 2006, p. 29.

Meat consumption levels of lions: Kathy Darling, *Lions*, Carolrhoda Books, Minneapolis, Minnesota, 2000, p. 22.

CHAPTER 6: "HOW SPEECH CAUSED THE SECOND GREAT WAVE OF HUMAN EVOLUTION—AND MIGHT EXPLAIN PREFERENTIAL HOMOSEXUALITY AND THE WAY WE RAISE OUR CHILDREN"

Robert Yerkes as a eugenicist: Robert W. Sussman, *The Myth of Race: The Troubling Persistence of an Unscientific Idea*, Harvard University Press, Cambridge, Massachusetts, 2014, p. 93.

Robert Yerkes' anti-Semitism: Joseph W. Bendersky, *The Jewish Threat: Anti-Semitic Politics of the U.S. Army*, Basic Books, New York, 2001, p. 168.

Brain extraction operations at the Yerkes National Primate Center: John O'Connor, "TV: 'Primate,' A Study by Wiseman," December 5, 1974, *The New York Times*, https://www.nytimes.com/1974/12/05/archives/tv-primate-a-study-by-wiseman-yerkes-research-unit-scrutinized-in.html.

Tool use by dolphins: Janet Mann and Brooke Sargent, "Tool Use in Wild Bottlenose Dolphins," *Encyclopedia of Marine Mammals*, 3rd ed., Bernd Wursig, J. G. M. Thewissen, Kit Kovacs, editors, Elsevier, Amsterdam, The Netherlands, November 2017, p. 1171.

Tool use by chinchillas: Robert W. Shumaker, Kristina R. Walkup, and Benjamin B. Beck, *Animal Tool Behavior: The Use and Manufacture of Tools by Animals*, 3rd ed., Johns Hopkins University Press, Baltimore, Maryland, 2011, p. ix.

Tool use by crows and other corvids: Jeremy Taylor, *Not A Chimp: The Hunt to Find the Genes That Make Us Human*, Oxford University Press, UK, 2007, pp. 183–216.

Jane Goodall's description of the aftermath of a chimp act of brutality, In the Shadow of Man: Goodall, p. 117.

Rate of ovine preferential homosexuality: John Schwartz, "Of Gay Sheep, Modern Science and Bad Publicity," *The New York Times*,

January 25, 2007, https://www.nytimes.com/2007/01/25/science/ 25sheep.html.

First account of ovine homosexuality: W. W. Swett, C. A. Matthews, and. R. R. Graves, "Early Recognition of the Freemartin Condition In Heifers Twin Born with Bulls," *Journal of Agricultural Research*, United States Department of Agriculture, July 1, 1940, vol. 41, no. 1, p. 587.

Chimerism among cows: M. D. Brace, O. Peters, P. Menzies, W. A. King, and M. I. Nino-Soto, "Sex chromosome chimerism and the freemartin syndrome in Rideau Arcott sheep," *Journal of Cytogenetics Genome Research*, e-published April 30, 2008, vol. 120, no. 1–2, pp. 132–139.

Causes of ovine homosexuality: Charles E. Roselli, Radhika Reddy, and Katherine Kaufman, "The Development of Male-Oriented Behavior in Rams," *Frontiers in Neuroendocrinology*, April 2011, vol. 32, no. 2, pp. 164–169.

Observable brain differences in homosexual men and women: Simon Levay, "A difference in hypothalamic structure between heterosexual and homosexual men," *Science* magazine, August 30, 1991, vol. 253, no. 5023, pp. 1034–1037. See also Amirhossein Manzouri and Ivanka Savic, "Multimodal MRI suggests that male homosexuality may be linked to cerebral midline structures," *PLOS One*, October 2, 2018, vol. 13, no. 10, https://journals.plos.org/plosone/article?id=10.1371/journal.pone.0203189. See also Mikhail Votinov, Katharina S. Goerlich, Andrei A. Puiu, Elke Smith, Thomas Nickl-Jockschat, Birgit Derntl, and Ute Habel, "Brain structure changes associated with sexual orientation," *Nature: Scientific Reports*, published online March 3, 2011, vol. 11, no. 5078, https://www.ncbi.nlm.nih.gov/pmc/articles/ PMC7930173/.

Homosexuality among herding species: Allen Rada, *Imposed Morality*, Inspiring Publishers, Calwell, Australia, 2019.

Relative frequency of exclusive male and female homosexuality: J. Michael Bailey, Paul L. Vasey, Lisa Diamond, Stephen Marc

Breedlove, Eric Vilain, and Marc Epprecht, "Sexual Orientation, Controversy, and Science," *Psychological Science in the Public Interest*, April 25, 2016, vol. 17, no. 2., pp. 45–101.

Relative frequency of male and female autism spectrum diagnoses: Rachel Loomes, Laura Hull, William Polmear Locke Mandy, "What Is the Male-to-Female Ratio in Autism Spectrum Disorder? A Systematic Review and Meta-Analysis," June 2017, *Journal of American Academy of Child and Adolescent Psychiatry*, vol. 56, pp. 466-474. At least four-fifths of autism diagnoses are of males. The authors believe the true case ratio is nearer to three to one. See also Simon Baron-Cohen, *The Essential Difference: the Male and Female Brains* , The Phi Beta Kappa Forum, 2005, as originally written January 1, 2003. https://www.researchgate.net/publication/23240614-The_Essential_Difference_The_Truth_About_The_Male_And_Female_Brain..

Role of the left occipito-temporal cortex in language processing: Quintino R. Mano, Colin Humphries, Rutvik H. Desai, Mark S. Seidenberg, David C. Osmon, Ben C. Stengel, and Jeffrey R. Binder, "The Role of Left Occipitotemporal Cortex in Reading: Reconciling Stimulus, Task, and Lexicality Effects," *Cerebral Cortex*, April 2013, vol. 23, no. 4, pp. 988–1001.

Machine learning analysis of observed differences in heterosexual and homosexual brains: Benjamin Clemens, Jeremy Lefort-Besnard, Christoph Ritter, Elke Smith, Mikhail Votinov, Birgit Derntl, Ute Habel, and Danilo Bzdok, "Accurate machine learning prediction of sexual orientation based on brain morphology and intrinsic functional connectivity," *Cerebral Cortex*, March 21, 2023, vol. 33, no. 7, pp. 4103–4025.

Observed chimp homosexual behavior: Goodall, *In the Shadow of Man*, pp. 183–184.

Homosexual behavior among bat species: Norimasa Sugita, "Homosexual Fellatio: Erect Penis Licking between Male Bonin Flying Foxes *Pteropus pselaphon*," *PLOS One*, November 8, 2016, vol. 11, no. 11, https://journals.plos.org/plosone/article?id=10.1371/journal.pone.0166024.

Relationship between primate group size and brain size: Rita Carter, *Mapping the Mind*, University of California Press, Ltd., London, UK, 1999, p. 157.

Interaction patterns of hunter-gatherer tribes: Kim R. Hill, Brian M. Wood, Jacopo Baggio, A. Magdalena Hurtado, and Robert T. Boyd, "Hunter-Gatherer Inter-Band Interaction Rates: Implications for Cumulative Culture," *PLOS One*, July 21, 2014, vol. 9, no. 7, https://www.ncbi.nlm.nih.gov/pmc/articles/PMC4105570/.

Eighteenth-century infant mortality rates: Irvine Loudon, "Deaths in childbed from the eighteenth century to 1935," *Medical History*, January 1986, vol. 30, no. 1, Cambridge University Press, UK, pp. 1–41.

Women display facial recognition better at earlier ages: Judith E. Owen Blakemore, Shari A. Berenbaum, and Lynn S. Liben, *Gender Development*, Psychology Press, New York, 2013.

Maternal mortality among gorillas and its effect on infant survival: Robin E. Morrison, Winnie Eckardt, Fernando Colchero, Veronica Vecellio, and Tara S. Stoinski, "Social groups buffer maternal loss in mountain gorillas," eLife, March 23, 2021, https://elifesciences.org/articles/62939. The study authors believe that this is because other members of a gorilla band will raise the gorilla as well as his mother would have. For this reason, it can be taken as evidence of gorilla sociability.

Growth of human brain in the first year of life: Ruth A. Lawrence and Robert M. Lawrence, "Breastfeeding," *Encyclopedia of Infant and Early Childhood Development*, 2nd ed., vol. 1:A–F, Jannette B. Benson, ed., Elsevier, Amsterdam, The Netherlands, 2020, p. 216.

Relative proportions of bodily growth in human infants: Jane B. Lancaster, "Human Adolescence and Reproduction: An Evolutionary Perspective," *School-Age Pregnancy and Parenthood: Biosocial Dimensions*, Jane B. Lancaster and Beatrix A. Hamburg, eds., Routledge, Milton Park, UK, 1986, pp. 22–23. See also Jane B.

Lancaster, "Sex Differences In Higher Primates," in *Gender and the Life Course*, edited by Alice Rossi, Aldine de Gruyter, Berlin, Germany, 1985, p. 20.

Difficulties of mastering a second language: Dana G. Smith, "At What Age Does Our Ability to Learn a New Language Like a Native Speaker Disappear?" *Scientific American*, May 4, 2018, https://www.scientificamerican.com/article/at-what-age-does-our-ability-to-learn-a-new-language-like-a-native-speaker-disappear/. See also Dominika Baran, *Language in Immigrant America*, Cambridge University Press, New York, 2017, p. 233.

Ants as a percentage of total jungle biomass: Edwin O. Wilson, *In Search of Nature*, Island Books, Shearwater Press, Washington, DC, 1996, p. 48.

Innate eusocial characteristics of bees: Martin A. Nowak, Corina E. Tarnita, and Edward O. Wilson, "The Evolution of Eusociality," *Nature*, August 26, 2010, vol. 466, no. 7310, pp. 1057–1062.

Wet-nursing observed among golden snub-nosed monkeys and other species: Zuofu Xiang, Penglai Fan, Haochun Chen, Ruoshuang Liu, Bo Zhang, Wanji Yang, Hui Yao, Cyril C. Grueter, Paul A. Garber, and Ming Li, "Routine allomaternal nursing in a free-ranging Old World monkey," *Science Advances*, February 20, 2019, vol. 5, no. 2. See also Sarah Blaffer Hrdy, *Mothers and Others: The Evolutionary Origins of Mutual Understanding*, Belknap Press, Cambridge, Massachusetts, 2011, p. 206.

CHAPTER 7: "HORMONES, HUMANS, HONEST SIGNALING, BIRDS, OCTOPUSES, AND CHIMPS"

Testosterone levels in healthy premenopausal women: University of Rochester Medical Center Health Encyclopedia, https://www.urmc.rochester.edu/encyclopedia/content.aspx?content typeid=167&contentid=testosterone_total.

Testosterone levels in postmenopausal women: Chevon M. Rariy, Sarah J. Ratcliffe, Rachel Weinstein, Shalender Bhasin, Marc R.

Blackman, Jane A. Cauley, John Robbins, Joseph M. Zmuda, Tamara B. Harris, and Anne R. Cappola, "Higher serum free testosterone concentration in older women is associated with greater bone mineral density, lean body mass, and total fat mass: the cardiovascular health study," *Journal of Clinical Endocrinology & Metabolism*, April 2011, vol. 96, no. 4, pp. 989–996.

Testosterone levels in chimp females: Victoria Wobber and Esther Herrmann, "The influence of testosterone on cognitive performance in bonobos and chimpanzees," *Behaviour*, May 7, 2014, vol. 152, p. 414, https://www.eva.mpg.de/documents/Brill/Wobber_Influence_Behaviour_2015_2076277.pdf.

Effects upon testosterone levels of fathers who spend time with their children: Jennifer Welsh, "A new study finds that levels of testosterone, the 'macho' sex hormone, drop in new fathers," *Scientific American*, September 12, 2011, https://www.scientificamerican.com/article/fatherhood-lowers-testosterone-keeps-dads-at-home/.

How human faces have changed over time: Karl Leif Bates, "Society Bloomed with Gentler Personalities and More Feminine Faces," Duke Today, August 1, 2014.

Observed effect of testosterone upon rate of correct answers to simple math questions: Gideon Nave, Amos Nadler, David Zava, and Colin Camerer, "Single-Dose Testosterone Administration Impairs Cognitive Reflection in Men," *Psychological Science*, August 3, 2017, vol. 28, no. 10, https://journals.sagepub.com/doi/10.1177/0956797617709592.

Likelihood of conception among healthy young heterosexual couples: Alison Taylor, "Extent of the Problem," *BMJ*, August 23, 2003, vol. 327, no. 7412, pp. 434–436, https://www.ncbi.nlm.nih.gov/pmc/articles/PMC188498/.

Oxytocin levels measured in chimps before and after fights: Liran Samuni, Anna Preis, Roger Mundry, Tobias Deschner, Catherine Crockford, and Roman M. Wittig, "Oxytocin reactivity during intergroup conflict in wild chimpanzees," *Proceedings of the*

National Academy of Sciences, January 10, 2017, vol. 114, no. 2, pp. 268–273, https://www.pnas.org/content/114/2/268.

Oxytocin and human impulses toward fighting: Hejing Zhang, Jörg Gross, Carsten De Dreu, and Yina Ma, "Oxytocin promotes coordinated out-group attack during intergroup conflict in humans," eLife, January 25, 2019, https://www.ncbi.nlm.nih.gov/pmc/articles/PMC6347450/.

Testosterone levels and instinct toward trusting others: "Peter A. Bos, David Terburg, and Jack van Honk, "Testosterone decreases trust in socially naïve humans," *Proceedings of the National Academy of Sciences*, May 24, 2010, vol. 107, no. 22, pp. 9991–9995, https://www.pnas.org/doi/10.1073/pnas.0911700107.

Comparison of DHEA levels in women and chimp females: Michael D. Gurven and Cristina M. Gomes, "Mortality, Senescence, and Life Span," in *Chimpanzees and Human Evolution*, edited by Martin N. Muller, Richard W. Wrangham, and David Pilbeam, Harvard University Press, Cambridge, Massachusetts, 2016, p. 198.

DHEA levels in primates: Benjamin Campbell, "DHEAS and Human Development: An Evolutionary Perspective," *Frontiers In Endocrinology*, March 3, 2020, vol. 11, no. 101, https://www.ncbi.nlm.nih.gov/pmc/articles/PMC7062714/.

Relationship between HAR1A and GABA production: Ke Xu, Eric E. Schadt, Katherine S. Pollard, Panos Roussos, and. Joel T. Dudley, "Genomic and network patterns of schizophrenia genetic variation in human evolutionary accelerated regions," *Molecular Biology and Evolution*, May 2015, vol. 32, no. 5, pp. 1148–60, e-published February 12, 2015. See also Bret Stetka, "Why Don't Animals Get Schizophrenia (and How Come We Do)?" *Scientific American*, March 24, 2015, https://www.scientificamerican.com/article/why-don-t-animals-get-schizophrenia-and-how-come-we-do/.

High levels of dopamine production observed among humans: Ann Gibbons, "Dopamine may have given humans our social edge over other apes," *Science*, January 22, 2018, https://www.science.org/content/article/dopamine-may-have-given-humans-our-social-edge-over-other-apes. See also Mary Ann Raghanti, Melissa K. Edler, Alexa R. Stephenson, Emily L. Munger, Bob Jacobs, Patrick R. Hof, Chet C. Sherwood, Ralph L. Holloway, and C. Owen Lovejoy, "A neurochemical hypothesis for the origin of hominids," *Proceedings of the National Academy of Sciences*, January 22, 2018, vol. 115, no. 6, pp. E1108–E1116.

GABA production among humans at differing ages and in both sexes: Jayarjun Ethiraj, Thulani Hansika Palpagama, Clinton Turner, Bert van der Werf, Henry John Waldvogel, Richard Lewis Maxwell Faull, and Andrea Kwakowsky, "The effect of age and sex on the expression of GABA signaling components in the human hippocampus and entorhinal cortex," *Scientific Reports*, November 2, 2021, vol. 11, no. 21470.

Apparent relationship between GABA production and autism spectrum disorders: Dirk Dhossche, Heather Applegate, Ann Abraham, Paul Maertens, Lorna Bland, Aladar Bencsath, and José Martinez, "Elevated plasma gamma-aminobutyric acid (GABA) levels in autistic youngsters: stimulus for a GABA hypothesis of autism," *Medical Science Monitor*, August 2002, vol. 8, no. 8, pp. 1–6.

Differing levels of key neurotransmitters in apes and humans: Ei Terasawa, "Role of GABA in the Mechanism of the Onset of Puberty in Non-Human Primates," *International Review in Neurobiology*, vol. 71, 2005, pp. 113–129.

Effect of EGF upon novelty-seeking: Mark J. Kwakkenbos, Mourad Matmati, Ole Madsen, Walter Pouwels, Yong Yi Wang, Ronald E. Bontrop, Peter J. Heidt, Robert M. Hoek, and Jörg Hamann, "An unusual mode of concerted evolution of the EGF-TM7 receptor chimera EMR2," *Federation of American Societies for Biological Sciences Journal*, October 26, 2006, vol. 20, no. 14, pp. 2582–2584. See also Liisa Keltikangas-Järvinen, Sampsa Puttonen, Minna Kivimäki, R. Rontu, and Terho Lehtimäki, "Cloninger's temperament dimensions and epidermal growth factor A61G polymorphism in Finnish adults," *Genes, Brain and Behavior*, 2006, vol. 5, no. 1, pp. 11–18.

Changes in the Y chromosome between humans and chimps: Jennifer F. Hughes, Helen Skaletsky, Tatyana Pyntikova, Tina A. Graves, Saskia K. M. van Daalen, Patrick J. Minx, Robert S. Fulton, Sean D. McGrath, Devin P. Locke, Cynthia Friedman, Barbara J. Trask, Elaine R. Mardis, Wesley C. Warren, Sjoerd Repping, Steve Rozen, Richard K. Wilson, and David C. Page, "Chimpanzee and human Y chromosomes are remarkably divergent in structure and gene content," *Nature*, January 28, 2010, vol. 463, no. 7280, pp. 536–539.

Effects of benzodiazepines on rodent behavior: Russell C. Leaf, D. J. Wnek, and Stacy Lamon, "Oxazepam induced mouse killing by rats," *Pharmacology Biochemistry and Behavior*, March 1984, vol. 20, no. 2, pp. 311–3. See also Russell C. Leaf, D. J. Wnek, Patricia E. Gay, R. M. Corcia, Russell C. Leaf, D. J. Wnek, Patricia E. Gay, R. M. Corcia, and Stacy Lamon, "Chlordiazepoxide and diazepam induced mouse killing by rats," *Psychopharmacology*, November 1975, vol. 44, no. 1, pp. 23–28.

Effects of amphetamines on rodent behavior: Patricia E. Gay, Russell C. Leaf, and Francine B. Arble, "Inhibitory effects of pre- and post-test drugs on mouse killing by rats," *Pharmacology, Biochemistry and Behavior*, February 1975, vol. 3, no. 1, pp. 33–45.

Chimp manipulativeness and calculation: De Waal, *Chimpanzee Politics*, p. 41.

Archeological evidence of the first recorded human burial sites: Daniella E. Bar-Yosef Mayer, Bernard Vandermeersch, and Ofer Bar-Yosef, "Shells and ochre in Middle Paleolithic Qafzeh Cave, Israel: indications for modern behavior," *Journal of Human Evolution*, March 2009, vol. 56, no. 3, pp. 307–314.

Elephants observed memorializing others: Martin Meredith, *Elephant Destiny: Biography of an Endangered Species in Africa*, PublicAffairs, New York, 2004, pp. 184–186. As quoted in article "Elephant Cognition" on Wikipedia, https://en.wikipedia.org/wiki/Elephant_cognition#cite_note-Meredith-42. See also

"Kenya Elephant Buries Its Victims," BBC News, June 18, 2004, http://news.bbc.co.uk/2/hi/africa/3818833.stm.

Dolphins observed acting to protect humans from sharks: Sam Jones, "Dolphins Save Swimmers from Shark," *The Guardian*, November 23, 2004. See also Margaret Gillrie-Fraser, *The Dolphin Way*, AuthorHouse, Bloomington, Indiana, 2011, p. 94.

Corvids observed memorializing others: Thom Van Dooren, *The Wake of Crows: Living and Dying in Shared Worlds,* Columbia University Press, New York, 2019.

Rats observed acting to protect one another and displaying empathy: Inbal Ben-Ami Bartal, Jean Decety, and Peggy Mason, "Empathy and Pro-Social Behavior in Rats," *Science*, December 9, 2011, vol. 334, no. 6061, pp. 1427–1430. See also Marc Bekoff, "Rats Share Food More Generously When They Smell Hunger," *Psychology Today*, March 28, 2020.

Patterns of chimp "pant-hooting": Alice Bouchard and Klaus Zuberbühler, "Male chimpanzees communicate to mediate competition and cooperation during feeding," *Animal Behaviour*, April 2022, vol. 186, pp. 41–55.

Frans de Waal tries to argue for chimp altruism and empathy by citing an example of canine altruism and empathy: De Waal, *Our Inner Ape*, pp. 180–181.

Infrequency of alloparenting among chimps: Iulia Bădescu, David P. Watts, M. Anne Katzenberg, and Daniel W. Sellen, "Alloparenting is associated with reduced maternal lactation effort and faster weaning in wild chimpanzees," *Royal Society Open Science*, October 12, 2016, https://royalsocietypublishing.org/doi/pdf/10.1098/rsos.160577.

Lack of observed active instruction by chimp and gorilla mothers with their offspring: Mia Sarkisian, "Orangutan Infant Behavior: A Critical Component for Primate Conservation," May 13, 2019, Bridgewater State University honors thesis. The author is referring to a 2002 study by Dario Maestripieri, Stephen K. Ross, and Nancy L. Megna, "Mother-Infant Interactions in Western

Lowland Gorillas (*Gorilla gorilla gorilla*): Spacial Relationships, Communication and Opportunities for Social Learning," *Journal of Comparative Psychology*, vol. 116, no. 3, pp. 219–227.

Increased foraging by grandmothers in hunter-gatherer tribes: Melissa Emery Thompson and Peter T. Ellison, "Fertility and Fecundity," in *Chimpanzees and Human Evolution*, edited by Martin N. Muller, Richard W. Wrangham, and David Pilbeam, Harvard University Press, Cambridge, Massachusetts, 2016, p. 242.

Relationship between alloparenting and altruism in primate species: J. M. Burkart, O. Allon, F. Amici, C. Fichtel, C. Finkenwirth, A. Heschl, J. Huber, K. Isler, Z. K. Kosonen, E. Martins, E. J. Meulman, R. Richiger, K. Rueth, B. Spillmann, S. Wiesendanger, and C. P. van Schaik, "The evolutionary origin of human hyper-cooperation," *Nature Communications*, August 27, 2014, vol. 5, no. 4714.

Paucity of observed costly signaling among chimps: Herbert Gintis, Eric Alden Smith, and Samuel Bowles, "Costly Signaling and Cooperation," *Journal of Theoretical Biology*, November 7, 2001, vol. 213, no. 1, pp. 114–115.

Principle of costly signaling explained: Francis T. McAndrew, "Costly Signaling Theory," *Encyclopedia of Evolutionary Psychological Science*, October 8, 2018.

Bird species known for costly signaling identified: Roni Ostreiher and Aviad Heifetz, "The sentinel behaviour of Arabian babbler floaters," *Royal Society Open Science*, February 1, 2017, vol. 4, no. 2.

FMRI research on canine altruism and empathy: Peter F. Cook, Ashley Prichard, Mark Spivak, and Gregory S. Berns, "Awake Canine fMRI Predicts Dogs' Preference for Praise Versus Food," *Social Cognitive and Affective Neuroscience*, August 2016, vol. 11, no. 12, pp. 1853–1862.

Manner in which chimps learn to groom one another: Richard W. Wrangham, Kathelijne Koops, Zarin P. Machanda, Steven Worthington, Andrew B. Bernard, Nicholas F. Brazeau, Ronan Donovan, Jeremiah Rosen, Claudia Wilke, Emily Otali, and Martin N. Muller, "Distribution of a Chimpanzee Social

Custom Is Explained by Matrilineal Relationship Rather Than Conformity," *Current Biology*, November 21, 2016, vol. 26, no. 22, pp. 3033–3037.

Practice of allogrooming among cows described: Inés de Freslon, J. M. Peralta, Ana C. Strappini, and Gustavo Monti, "Understanding Allogrooming Through a Dynamic Social Network Approach: An Example in a Group of Dairy Cows," *Frontiers in Veterinary Science*, August 4, 2020, vol. 7, https://doi.org/10.3389/fvets.2020.00535.

Motives of chimps grooming other chimps: Nicholas E. Newton-Fisher and Stefano S. K. Kaburu, "Grooming decisions under structural despotism: the impact of social rank and bystanders among wild male chimpanzees," *Animal Behaviour*, June 2017, vol. 128, pp. 153–164. See also Goodall, *In the Shadow of Humans*, pp. 190–191.

Suicide among incarcerated inmates: Morris L. Thigpen, Thomas J. Beauclair, Virginia A. Hutchinson, and Fran Zandi, "National Study of Jail Suicide: 20 Years Later," US Department of Justice, National Institute of Corrections, Washington, DC, 2006, pp. 15–19, 22–23, and 27. In the US for those in holding facilities, 45 percent of suicides take place within the first six hours. Roughly two-thirds kill themselves with their bedsheets, and 92 percent kill themselves by hanging.

Effect of forced isolation on a captive gorilla: Laurel Braitman, *Animal Madness: How Anxious Dogs, Compulsive Parrots, and Elephants in Recovery Help Us Understand Ourselves*, Simon & Schuster, New York, 2014, pp. 173–174.

Animals that can pass a mirror test: Amanda Pachniewska, "List of Animals That Have Passed the Mirror Test," *Animal Cognition*, April 15, 2015, https://www.animalcognition.org/2015/04/15/list-of-animals-that-have-passed-the-mirror-test/.

Report of unusual skill of a parrot with human language: Alex Kirby, "Parrot's Oratory Stuns Scientists," BBC News, January 26, 2004, http://danielwebb.us/opinion/web_journal/bbc_bad_science/telepathic_bird/3430481.stm.html.

Innate fear of snakes observed among infants: Paul Bloom, *Descartes' Baby: How the Science of Child Development Explains What Makes Us Human*, Basic Books, New York, 2005, p. 223.

Innate fear of cats observed among rats: Virginia Morell, *Animal Wise: How We Know Animals Think and Feel*, Broadway Books, New York, 2014, p. 117.

Filial cannibalism among chimps and bonobos reported: Pawel Fedurek, Patrick Tkaczynski, Caroline Asiimwe, Catherine Hobaiter, Liran Samuni, Adriana E. Lowe, Appolinaire Gnahe Dijrian, Klaus Zuberbühler, Roman M. Wittig, and Catherine Crockford, "Maternal cannibalism in two populations of wild chimpanzees," *Primates*, October 5, 2019, vol. 61, pp. 181–187. See also Ewen Callaway, "Hippy apes caught cannibalising their young," *New Scientist*, February 1, 2010, https://www.newscientist.com/article/dn18454-hippy-apes-caught-cannibalising-their-young/.

Heavy meat consumption and projected life expectancy: Michael J. Orlich, Pramil N. Singh, Joan Sabaté, Karen Jaceldo-Siegl, Jing Fan, Synnove Knutsen, W. Lawrence Beeson, and Gary E. Fraser, "Vegetarian dietary patterns and mortality in Adventist Health Study 2," *JAMA Internal Medicine*, July 8, 2013, vol. 173, no. 13, pp. 1230–1238.

Chimp proficiency at breaking up nut shell casings: Jeffrey K. McKee, Frank E. Poirier, and W. Scott McGraw, *Understanding Human Evolution*, 5th ed., Routledge, Milton Park, UK, 2016, p. 94.

Report of a chimp kidnapping and eating a human baby: De Waal, *Our Inner Ape*, pp. 137–138.

Reports of chimps eating baboon infants: Goodall, *In the Shadow of Man*, pp. 194–204.

Chimp fertility patterns: Melissa Emery Thompson, James H. Jones, Anne E. Pusey, Stella Brewer-Marsden, Jane Goodall, David Marsden, Tetsuro Matsuzawa, Toshisada Nishida, Vernon Reynolds, Yukimaru Sugiyama, and Richard W. Wrangham,

"Aging and fertility patterns in wild chimpanzees provide insights into the evolution of menopause," *Current Biology*, December 18, 2007, vol. 17, no. 24, pp. 2150–2156. See also Yukimaru Sugiyama, "Demographic Parameters and Life History of Chimpanzees at Bossou, Guinea," *American Journal of Physical Anthropology*, vol. 124, 2004, pp. 154–165.

Human and dolphin fertility patterns: Nicholas Blurton-Jones, "Bushman birth spacing: A test for optimal interbirth intervals," *Ethology and Sociobiology*, 1986, vol. 7, no. 2, pp. 91–105.

CHAPTER 8: "DO YOU THINK HUMANS ARE SEXY? HERE'S WHY YOU DO"

High fertility rates of older female chimpanzees: Melissa Emery Thompson, James H. Jones, Anne E. Pusey, Stella Brewer-Marsden, Jane Goodall, David Marsden, Tetsuro Matsuzawa, Toshisada Nishida, Vernon Reynolds, Yukimaru Sugiyama, and Richard W. Wrangham, "Aging and fertility patterns in wild chimpanzees provide insights into the evolution of menopause," *Current Biology*, December 18, 2007, vol. 17, no. 24, pp. 2150–2156. See also Goodall, *In the Shadow of Man*, pp. 81–86.

Lack of reported ovulation synchrony among female primates: Akiko Matsumoto-Oda, Miya Hamai, Hitosige Hayaki, Kazuhiko Hosaka, Kevin D. Hunt, Eiiti Kasuya, Kenji Kawanaka, John C. Mitani, Hiroyuki Takasaki, and Yukio Takahata, "Estrus cycle asynchrony in wild female chimpanzees, *Pan troglodytes schweinfurthii*," *Behavioral Ecology and Sociobiology*, 2007, vol. 61, pp. 661–668. See also Joanna M. Setchell, Jeremy Kendal, and Pola Tyniec, "Do non-human primates synchronise their menstrual cycles? A test in mandrills," *Psychoneuroendocrinology*, January 2011, vol. 36, no. 1, pp. 51–59.

Reported ovulation synchrony among female elephants: Nicole M. Weissenböck, Harald M. Schwammer, and Thomas Ruf, "Estrous synchrony in a group of African elephants (*Loxodonta africana*) under human care," *Animal Reproduction Science*, July 2009, vol. 113, no.1–4, pp. 322–327.

Relationship between height and attractiveness in men: Gert Stulp, Abraham P. Buunk, Robert Kurzban, and Simon Verhulst, "The height of choosiness: mutual mate choice for stature results in suboptimal pair formation for both sexes," *Animal Behaviour*, 2013, vol. 86, no. 1, pp. 37–46. On average, women indicated that the ideal male height is 25.1 cm. greater than their own height. This is 9.9 inches. Since the average woman is 5'4" inches tall, ideal male height, according to women, is about 6'2".

Maximum speed of gibbons when swinging on tree branches: Ewen Callaway, "How gibbons got their swing," *Nature*, September 11, 2014, https://www.nature.com/articles/nature.2014.15895.

Diet and habitat of gibbons: Pandey and Mathur, *Biology of Chordates*, p. 504.

Inverse relationship between the penis bone and monogamous mating patterns: Carl Zimmer, "A Most Interesting Bone," *National Geographic*, August 8, 2013, https://www.nationalgeographic.com/science/article/a-most-interesting-bone?loggedin=true.

Relationship between feminine figures and fertility: Grazyna Jasieńska, Anna Ziomkiewicz, Peter T. Ellison, Susan F. Lipson, and Inger Thune, "Large breasts and narrow waists indicate high reproductive potential in women," *Proceedings of the Royal Society B: Biological Sciences*, June 22, 2004, vol. 271, no. 1545, pp. 1213–1217.

High body-fat ratios of Japanese snow monkeys: Tomoko Takahashi, Atsunori Higashino, Kaori Takagi, Yoshiro Kamanaka, Masamitsu Abe, Mayumi Morimoto, Kyung Hwa Kang, Shunji Goto, Juri Suzuki, Yuzuru Hamada, and Takashi Kageyama, "Characterization of obesity in Japanese monkeys (*Macaca fuscata*) in a pedigreed colony," *Journal of Medical Primatology*, February 2006, vol. 35, no. 1, pp. 30–37.

Low body-fat ratios of baboons: High vocal register of chimp males: Martin N. Muller and David R. Pilbeam, "The Evolution of the

Human Mating System," in *Chimpanzees and Human Evolution*, edited by Martin N. Muller, Richard W. Wrangham, and David Pilbeam, Harvard University Press, Cambridge, Massachusetts, 2016, p. 396.

High vocal register of chimp males: Kalan K. Ammie, "Evidence for sexual dimorphism in chimpanzee vocalizations: a comparison of male and female call production and acoustic parameters," in *The Chimpanzees of the Taï Forest: 40 Years of Research*, edited by Christophe Boesch and Roman Wittig, Cambridge University Press, UK, published online by Cambridge University Press, November 25, 2019.

Relationship between human male voice registers and healthy male immune function: David A. Puts, Alexander K. Hill, Drew H. Bailey, Robert S. Walker, Drew Rendall, John R. Wheatley, Lisa L. M. Welling, Khytam Dawood, Rodrigo Cárdenas, Robert P. Burriss, Nina G. Jablonski, Mark D. Shriver, Daniel Weiss, Adriano R. Lameira, Coren L. Apicella, Michael J. Owren, Claudia Barelli, Mary E. Glenn, and Gabriel Ramos-Fernandez, "Sexual selection on male vocal fundamental frequency in humans and other anthropoids," *Proceedings of the Royal Society B: Biological Sciences*, April 27, 2016, vol. 283, no. 1829.

Frans de Waal claims that chimps have a similar sense of fairness as humans: De Waal, *Mama's Last Hug*, p. 218.

Murder of baboon infants by rival males: "Inside Chacma Baboon Society," *Nature*, PBS, May 30, 2008.

Theft of baboon infants by rival mothers: "European Zoo and Aquaria Best Practices Guidelines," Gelada baboon (*Theropithecus gelada*), p. 47, https://strapi.eaza.net/uploads/Gelada_baboon_BPG_final_edit_070521_compressed_993327afe8.pdf.

Lack of instruction of young baboons by their mothers: Jeanne Altmann, *Baboon Mothers and Infants*, University of Chicago Press, 1980, Chicago, p. 130.

Superior skill of rooks at problem-solving: Christopher D. Bird and Nathan J. Emery, "Rooks perceive support relations similar to six-month-old babies," *Proceedings of the Royal Society B: Biological Sciences*, January 7, 2010, vol. 277, no. 1678, pp. 147–151.

Concept of social monogamy explained: U. H. Reichard, "Monogamy: Past and present," in *Monogamy: Mating Strategies and Partnerships in Birds, Humans, and Other Mammals*, edited by U. H. Reichard and Christophe Boesch, Cambridge University Press, UK, 2003, pp. 3–25 as quoted in Christopher Ryan and Cacilda Jethá, "But Honey, I Thought You Meant 'Socially' Monogamous!" *Psychology Today*, September 19, 2013.

Claims about benefits of polyamory in Northeastern Quebec: Christopher Ryan and Cacilda Jethá, *Sex at Dawn: How We Mate, Why We Stray, and What It Means for Modern Relationships*, Harper Perennial, New York, 2010, p. 107.

Child mortality rates among the Inuit of Northeastern Quebec: Kathryn A. Wotton, "Mortality of Labrador Innu and Inuit, 1971–1982," *Circumpolar Health*, vol. 84, pp. 139–142.

Effects of polygamy on child rearing: Deborah T. Cragun and Ryan T. Cragun, "The Intricacies and Ethics of Parental Genetic Testing," *Modern Polygamy in the United States*, edited by Cardell Jacobson and Lara Burton, Oxford University Press, Oxford, New York, 2011, p. 336.

Effects of polygamy on child mortality in Mali: David Barash, *Out of Eden: The Surprising Consequences of Polygamy*, Oxford University Press, New York, 2016, pp. 88–90.

Effects of polygamy in nineteenth-century Utah: Paul W. Reeve, *Religion of a Different Color: Race and the Mormon Struggle For Whiteness*, Oxford University Press, New York, 2015, p. 15.

The crimes of Warren Jeffs, Mormon polygamist: Alex Hannaford, "The evil preacher who runs his cult from prison," *The Daily Telegraph*, June 17, 2014, http://www.telegraph.co.uk/culture/tvandradio/10904019/The-evil-preacher-who-runs-his-cult-from-prison.html.

Increased likelihood of sexual molestation by non-biological fathers: Diana E. H. Russell, "The prevalence and seriousness of incestuous abuse: Stepfathers vs. biological fathers," *Journal of Child Abuse and Neglect*, 1984, vol. 8, no. 1, pp. 5–22.

Infrequency of false paternity in human populations: J. M. Greeff and J. C. Erasmus, "Three hundred years of low non-paternity in a human population," *Heredity*, May 6, 2015, vol. 115, pp. 396–404. See also Kermyt G. Anderson, "How Well Does Paternity Confidence Match Actual Paternity? Evidence from Worldwide Nonpaternity Rates," *Current Anthropology*, vol. 47, no. 3. See also Razib Khan, "The Paternity Myth: The Rarity of Cuckoldry," *Discover*, June 20, 2010. See also Carl Zimmer, "Fathered by the Mailman? It's Mostly an Urban Legend," *The New York Times*, April 8, 2016. See also Maarten H. D. Larmuseau, Pieter van den Berg, Sofie Claerhout, Kelly Nivelle, Ronny Decorte, and Tom Wenseleers, "A Historical-Genetic Reconstruction of Human Extra-Pair Paternity," *Current Biology*, December 2, 2019, vol. 29, no. 23, pp. 4102–4107.

Frequency of false paternity in pair-bonding bird species: Lyanne Brouwer and Simon C. Griffith, "Extra-pair paternity in birds," *Molecular Ecology*, November 2019, vol. 28, no. 22, pp. 4864-4882.

Influence of social structure on rates of false paternity: Aurélie Cohas and Dominique Allainé, "Social structure influences extra-pair paternity in socially monogamous mammals," *Biology Letters*, June 23, 2009, vol. 5, no. 3, pp. 313–316.

Relationship between life expectancy and fidelity in monogamous bird species: Simon C. Griffith, Ian P. F. Owens, and Katherine A. Thuman, "Extra pair paternity in birds: a review of interspecific variation and adaptive function," *Molecular Ecology*, November 2002, vol. 11, no. 11, pp. 2195–2212.

Rates of divorce reported as higher among lesbians than gay men: Nick Duffy, "Lesbians significantly more likely to get divorced than gay men, according to science," *PinkNews* (UK), November 18, 2020, https://www.pinknews.co.uk/2020/11/18/married-lesbi-an-couples-divorce-gay-men-office-national-statistrics-data/. See also Martin Kolk and Gunnar Andersson, "Two Decades of Same-Sex Marriage in Sweden: A Demographic Account of Developments in Marriage, Childbearing, and Divorce," *Demography*, 2020, vol. 57, pp. 147–169. See also Janene Pieters, "Marriages between Women Most Likely to End in Divorce," *NL Times* (The Netherlands), March 1, 2016. See also Sarah Prager, "First Comes U-Haul, Then Comes Court: The Shockingly High Lesbian Divorce Rate," GOMag.com, February 19, 2020.

Frequency of sexual intercourse among heterosexual married couples: Jerrold S. Greenberg, Clint E. Bruess, and Sarah C. Conklin, *Exploring the Dimensions of Human Sexuality*, 3rd ed., Jones & Bartlett Learning, Burlington, Massachusetts, 2007, p. 429.

New York Times points out some benefits of monogamous parenting: Carl Zimmer, "Monogamy and Human Evolution," *The New York Times*, August 2, 2013.

Ryan and Jethá criticize The New York Times: Ryan and Jethá, "But Honey, I Thought You Meant 'Socially' Monogamous!"

Causes of homicide identified: Robert Charles Levie, "A Study of Criminal Homicide Patterns and Selected Characteristics of Victims and Offenders in the City of New Orleans, 1971–73," Louisiana State University doctoral dissertation, 1979, p. 152.

Causes of depression and anxiety among prostitutes: Young-Eun Jung, Jeong-Min Song, Jihye Chong, Ho-Jun Seo, and Jeong-Ho Chae, "Symptoms of Posttraumatic Stress Disorder and Mental Health in Women Who Escaped Prostitution and Helping Activists in Shelters," *Yonsei Medical Journal*, June 30, 2008, vol. 49, no. 3, pp. 372–82.

CHAPTER 9: "WHY PEOPLE COMMIT MURDERS, START WARS, AND BUILD PYRAMIDS"

Comparative rates of murder among animals: Joseph Castro, "Do Animals Murder Each Other?" Live Science, September 16, 2017, https://www.livescience.com/60431-do-animals-murder-each-other.html.

Homicide rates by country: Mark Pagel, "Animal behavior: Lethal violence deep in the human lineage," *Nature*, October 13, 2016, vol. 538, pp. 180–181. See also "Murder Rate by Country, 2021," https://worldpopulationreview.com/country-rankings/murder-rate-by-country. See also Maia Szalavitz, "The surprising factors driving murder rates: income inequality and respect," *The Guardian*, December 8, 2017. See also Barash, *Out of Eden*, p. 49.

Archeological evidence of Phoenician child sacrifice: Katelyn DiBenedetto, "Analyzing Tophets: Did the Phoenicians Practice Child Sacrifice?," Anthropology Scholars Archive, State University of New York at Albany, honors thesis, March 2012, pp. 2–32.

Annual numbers of ritual animal sacrifices in Pakistan: "Pakistanis to sacrifice over 10 million animals this Eid," GeoTV, September 12, 2016, https://www.geo.tv/latest/114495-Pakistanis-to-sacrifice-over-10-million-animals-this-Eid.

Cannibalism following overthrow of alpha male chimps observed: Ilana Gordon, "Chimps Beat Up, Murder and Eat Their Tyrant," Medium, February 1, 2017.

Kidnapping of siblings by siblings as a means of manipulating matriarchs observed among chimps: Goodall, *In the Shadow of Man*, pp. 108–109.

Tamerlane's death toll: Tucker, *Marriage and Civilization*, p. 173.

Monthly cost of the Ayatollah Khomeini's primary residence: Lee Stokes, "Khomeini's home opened for tour," UPI Archives, June 7, 1989. See also Iranian Tours: The House of Khomeini, https://www.iraniantours.com/attraction/the-house-of-imam-khomeini/.

Pyramids were most likely not built by slaves: Jonathan Shaw, "Who Built the Pyramids? Not slaves. Archaeologist Mark Lehner, digging deeper, discovers a city of privileged workers," *Harvard Magazine*, July–August 2003. https://www.harvardmagazine.com/2003/07/who-built-the-pyramids-html.

Higher intelligence of wolves than dogs claimed: Michelle Lampe, Juliane Bräuer, Juliane Kaminski, and Zsófia Virányi, "The effects of domestication and ontogeny on cognition in dogs and wolves," *Scientific Reports*, September 15, 2017, vol. 7, no. 11690.

Worldwide practice of arranged marriage: Gwen J. Broude and Sarah J. Greene, "Cross-cultural Codes on Husband-Wife Relationships," *Journal of Ethnology*, July 2003, vol. 22, no. 3, pp. 263–280, as quoted in Wikipedia article on "Arranged Marriages."

Frequency and unwelcome health effects of consanguineous marriage in the Middle East: Sharkia Rajech, Athamny Esmael, Khatib Mohamad, Sheikh-Muhammad Ahmad, Azem Abdussalam and Mahajnah Muhammad "Consanguinity and Its Effect on Morbidity and Congenital Disorders Among Arabs In Israel," *Human Genetic Diseases*, edited by Dijana Plaseska-Karanfilska, *InTech*, Rijeka, Croatia, 2011, p. 271–274.

Turkish government promotes polygamy: Pinar Tremblay, "Erdogan's silent revolution seeks more than pious youth," Al-Monitor, June 17, 2021.

Adolf Hitler wished to allow polygamy for decorated German war veterans: Paul Johnson, *Modern Times: From the Twenties to the Nineties*, rev. ed., HarperCollins, New York, 1991, p. 381.

Wide distribution of genes believed to descend from Genghis Khan: Tatiana Zerjal, Yali Xue, Giorgio Bertorelle, R. Spencer Wells, Weidong Bao, Suling Zhu, Raheel Qamar, Qasim Ayub, Aisha Mohyuddin, Songbin Fu, Pu Li, Nadira Yuldasheva, Ruslan Ruzibakiev, Jiujin Xu, Qunfang Shu, Ruofu Du, Huanming Yang, Matthew E. Hurles, Elizabeth Robinson, Tudevdagva Gerelsaikhan, Bumbein Dashnyam, S. Qasim Mehdi, and Chris Tyler-Smith, "The Genetic Legacy of the Mongols," *American Journal of Human Genetics*, March 2003, vol. 72, no. 3, pp. 717–721.

Results of gene tests of Napoleon's claimed and actual descendants: Gerard Lucotte and Peter Hrechdakian, "New Advances Reconstructing the Y Chromosome Haplotype of Napoleon the First Based on Three of His Living Descendants," *Journal of Molecular Biology*

Research, December 19, 2014, vol. 5, no. 1., https://www.ccsenet. org/journal/index.php/jmbr/article/view/40064.

Number of America's richest individuals who have pledged half their fortune to charity: Mattathias Schwartz, "Bill Gates and Warren Buffett got 211 billionaires to pledge half their wealth to charity. But some are moving slow—and still getting massive tax breaks," Business Insider, May 20, 2021, https://www.businessinsider. com/giving-pledge-bill-gates-warren-buffettt-billionaires-tax-break-2021-5.

CHAPTER 10: "MAN, THE UNTHINKING ANIMAL"

Observed effects of the influence of reported taste on stated musical preference: Matthew J. Salganik, Peter Sheridan Dodds, and Duncan J. Watts, "Experimental Study of Inequality and Unpredictability in an Artificial Cultural Market," *Science*, February 10, 2006, vol. 311, pp. 854–856.

Data on the tulipmania phenomenon in seventeenth-century Holland: Alastair Sooke, "Tulip mania: The flowers that cost more than houses," BBC, May 3, 2016.

The extreme effects of Japan's land boom in the 1980s: Edward Jay Epstein, "What Was Lost (and Found) In Japan's Lost Decade," *Vanity Fair*, February 17, 2009.

Inhumane practices of modern industrial hog farms: Erik Marcus, *Vegan: The New Ethics of Eating*, rev. ed., McBooks Press, Ithaca, New York, 2001, p. 120.

High mortality rates of African-descended slaves in the New World: Steven Mintz, "History Resources—Historical Context: Facts About the Slave Trade and Slavery," Gilder Lehrman Institute of American History, https://www.gilderlehrman.org/history-resources/teaching-resource/historical-context-facts-about-slave-trade-and-slavery.

Effects of social rejection parallel physical pain: Naomi I. Eisenberger, "The neural bases of social pain: Evidence for shared representations with physical pain," *Psychosomatic Medicine*, February 2012, vol. 74., no. 2, pp. 126–135.

CHAPTER 11: "WHY WE CAN RUN MARATHONS BUT NOT OUTRUN BEARS"

Chimp males use meat to attract females: Cristina M. Gomes and Christophe Boesch, "Wild Chimpanzees Exchange Meat for Sex on a Long-Term Basis," *PLOS One*, e-published April 8, 2009, vol. 4, no. 4., https://www.ncbi.nlm.nih.gov/pmc/articles/PMC2663035/.

Dangerousness of chimps as depicted by Frans de Waal: De Waal, *Chimpanzee Politics*, p. 14.

Reported effects of captivity upon the behavior of captive animal populations: M. E. Mcphee and Kathy Carlstead, "Effects of Captivity on the Behavior of Wild Mammals," in *Wild Mammals in Captivity*, January 2010, University of Chicago Press, Chicago, pp. 303–313.

Rapid changes upon the genome of captive animal populations reported: Mark R. Christie, Melanie L. Marine, Rod A. French, and Michael S. Blouin, "Genetic adaptation to captivity can occur in a single generation," *Proceedings of the National Academy of Sciences*, January 3, 2012, vol. 109, no. 1, pp. 238–242, https://doi.org/10.1073/pnas.1111073109.

Average height of male pygmies: Luigi Luca Cavalli-Sforza, Paolo Menozzi, and Alberto Piazza, *The History and Geography of Human Genes*, abridged paperback ed., Princeton University Press, Princeton, New Jersey, 2018, p. 167.

Cooperation among women and children in hunting of okapis: David S. Wilkie and Bryan Curran, "Why Do Mbuti Hunters Use Nets? Hunting Efficiency of Archers and Net-Hunters in the Ituri Rain Forest," *American Anthropologist*, September 1991, vol. 93, no. 3, pp. 680–689.

Animals reported to be best at activities requiring endurance: Brian Resnick, "The Animal Kingdom's Top Marathoners," *Popular*

Mechanics, November 5, 2010. See also Peter Davis Krahenbuhl, "Gray Whale—Greatest Endurance Athlete in the World?" GoodNature.com, November 21, 2016, https://www.nathab.com/blog/gray-whale-greatest-endurance-athlete-in-the-world/.

Human skill at sweating when compared with chimps: Daniel Aldea, Yuji Atsuta, Blerina Kokalari, and Yana G. Kamberov, "Repeated mutation of a developmental enhancer contributed to human thermoregulatory evolution," *Proceedings of the National Academy of Sciences: Biological Sciences*, April 13, 2021, vol. 118, https://www.pnas.org/doi/full/10.1073/pnas.2021722118.

Anatomical differences between human and ape hearts: Robert E. Shave, Daniel E. Lieberman, Aimee L. Drane, Marcel G. Brown, Alan M. Batterham, Steven Worthington, Rebeca Atencia, Yedra Feltrer, Jennifer Neary, Rory B. Weiner, Meagan M. Wasfy, and Aaron L. Baggish, "Selection of endurance capabilities and the trade-off between pressure and volume in the evolution of the human heart," *Proceedings of the National Academy of Sciences*, October 1, 2019, vol. 116, no. 40, pp. 19905–19910.

Average age of death and early onset of sexual activity of wild chimps: Steve Bradt, "Chimps in wild appear not to regularly experience menopause," *Harvard Gazette*, December 13, 2007, https://news.harvard.edu/gazette/story/2007/12/chimps-in-wild-appear-not-to-regularly-experience-menopause/.

Relationship between tameness and paedomorphism: Richard Wrangham, *The Goodness Paradox: The Strange Relationship Between Virtue and Violence in Human Evolution*, Knopf Doubleday Publishing Group, New York, 2019, p. 174.

Differences observed between human and Neanderthal dentition: Tanya M. Smith, Paul Tafforeau, Donald J. Reid, Joane Pouech, Vincent Lazzari, John P. Zermeno, Debbie Guatelli-Steinberg, Anthony J. Olejniczak, Almut Hoffman, Jakov Radovčić, Masrour Makaremi, Michel Toussaint, Chris Stringer, and Jean-Jacques Hublin, "Dental evidence for ontogenetic differences between modern humans and Neanderthals," *Proceedings of the National*

Academy of Sciences, December 7, 2010, vol. 107, no. 49, https://www.pnas.org/content/107/49/20923.

Comparison of human and ape metabolism: Ann Gibbons, "The calorie counter: Evolutionary anthropologist Herman Pontzer busts myths about how humans burn calories—and why," *Science*, February 17, 2022, vol. 375, no. 6582, www.science.com/content/article/scientist-busts-myths-about-how-humans-burn-calories-and-why.

Higher metabolic rates of carnivorous animals observed: Agustí Muñoz-Garcia and Joseph B. Williams, "Basal Metabolic Rate in Carnivores Is Associated with Diet after Controlling for Phylogeny," *Physiological and Biochemical Zoology*, vol. 78, no. 6.

CHAPTER 12: "SENSE, SENSIBILITY, AND SPOKEN LANGUAGE"

Study showing more observant goats are less popular within the herd: Alvaro L. Caicoya, Alina Schaffer, Ruben Holland, Lorenzo von Fersen, Montserrat Colell, and Federica Amici, "Innovation across 13 ungulate species: problem solvers are less integrated in the social group and less neophobic," *Proceedings of the Royal Academy of Sciences B: Biological Sciences*, April 5, 2023, https://doi.org/10.1098/rspb.2022.2384.

Exceptionally poor concentration and task completion shown in studies comparing chimps with other animals: Johan Lind, Magnus Enquist, and Stefano Ghirlanda, "Animal memory: A review of delayed matching-to-sample data," *Journal of Behavioural Processes*, August 2015, vol. 117, pp. 52–58. See also Katharina C. Kirchhofer, Felizitas Zimmermann, Juliane Kaminski, and Michael Tomasello, "Dogs (*Canis familiaris*), but Not Chimpanzees (*Pan troglodytes*), Understand Imperative Pointing," *PLOS One*, February 8, 2012, vol. 7, no. 2, https://journals.plos.org/plosone/article?id=10.1371/journal.pone.0030913. See also Evan L. MacLean, Esther Herrmann, Sunil Suchindran, and Brian Hare, "Individual differences in cooperative communicative skills are more similar between dogs and humans than chimpanzees," *Animal Behaviour,* April 2017, vol. 126, no. 2, pp. 41–51.

Chimps score lower on intelligence tests than rhesus monkeys: De Waal, *Mama's Last Hug,* p. 80.

Study reports chimp disinclination to watch faces: Masako Myowa-Yamakoshi, Chisato Yoshida, and Satoshi Hirata, "Humans but Not Chimpanzees Vary Face Scanning Patterns Depending on Contexts During Action Observation," *PLOS One,* vol. 10, no. 11, November 4, 2015, https://europepmc.org/backend/ptpmcrender.fcgi?accid=PMC4633149&blobtype=pdf.

Paralimbic region of the dolphin brain noted: Lori Marino, "The Brain: Evolution, Structure and Function," in *Dolphin Communication: Past, Present and Future,* edited by Christine M. Johnson and Denise L. Herzing, MIT Press, Cambridge, Massachusetts, September 2015, p. 10.

Oral communication observed among dolphins during activities requiring joint task completion: Holli C. Eskelinen, Kelley A. Winship, Brittany L. Jones, Audra E. M. Ames, and Stan A. Kuczaj II, "Acoustic behavior associated with cooperative task success in bottlenose dolphins (*Tursiops truncatus*)," *Animal Cognition,* March 29, 2016, vol. 19, pp. 789–797. See also V. M. Janik, L. S. Sayigh, and R. S. Wells, "Signature whistle shape conveys identity information to bottlenose dolphins," *Proceedings of the National Academy of Sciences,* May 23, 2006, vol. 103, no. 21, pp. 8293–8297. See also Hal Hodson, "Dolphin whistle instantly translated by computer," March 26, 2014, NewScientist.com, https://www.newscientist.com/article/mg22129624-300-dolphin-whistle-instantly-translated-by-computer/.

Causes of dolphin menopause examined: David P. Barash, *Homo Mysterious: Evolutionary Puzzles of Human Nature,* Oxford University Press, New York, 2012, pp. 80–83.

Researchers suggest that dolphins are more like humans than apes: Andrew Whiten, "Imitation and Cultural Transmission in Apes and Cetaceans," *Behavioral and Brain Sciences*, 2001, vol. 24, no. 2, pp. 359–360. See also Lori Marino, R. C. Connor, R. E. Fordyce, L. M. Herman, P. R. Hof, L. Lefebvre, D. Lusseau, B. McCowan, E. A. Nimchinsky, A. A. Pack, L. Rendell, J. S. Reidenberg, D. Reiss, M. D. Uhen, E. Van der Gucht, and H. Whitehead, "Cetaceans have complex brains for complex cognition," *PLOS Biology*, vol. 5, no. 5, e139, pp. 966–972, https://doi.org/10.1371/journal.pbio.0050139.

Orangutan calls are not meant to encourage mutuality: Jean-Jacques Petter and François Desbordes, *Primates: An Illustrated Guide*, Princeton University Press, Princeton, New Jersey, 2013, p. 30.

CHAPTER 13: "CAUSES OF SUICIDE"

Suicide rates by sex, age, and country reviewed: "Suicide Rate by Country 2025," World Population Review, https://worldpopulationreview.com/country-rankings/suicide-rate-by-country. See also "Suicide in Japan," Wikipedia, https://en.wikipedia.org/wiki/Suicide_in_Japan#cite_note-Chambers-9.

Frequency and antiquity of reports of cetacean beaching and stranding: Rain Jordan, "Cetacean Strandings: Whales Are Beaching Themselves, Who Do We Blame?" *Nature World News*, updated March 30, 2021, https://www.natureworldnews.com/articles/45573/20210330/cetacean-strandings-whales-beaching-themselves-who-blame.htm. See also Braitman, *Animal Madness*, p. 167.

Toxic algae as an explanation for sea lion acts of self-destruction: "Toxic Algae Causing Brain Damage in Sea Lions Along California Coast," *Scientific American,* December 16, 2015, https://www.scientificamerican.com/article/toxic-algae-causing-brain-damage-in-sea-lions-along-california-coast/.

Men more likely than women to commit suicide: "Suicide Statistics," American Foundation for Suicide Prevention, https://afsp.org/suicide-statistics/.

Childbirth as a prophylactic against suicide: Chun-Yuh Yang, "Association between parity and risk of suicide among parous women," *Canadian Medical Association Journal*, April 4, 2010, vol. 182, no. 6, https://www.ncbi.nlm.nih.gov/pmc/articles/PMC2845684/.

Stabilizing effects of prolactin release: Markus Heinrichs, Gunther Meinlschmidt, Inga Neumann, Sabine Wagner, Clemens Kirschbaum, Ulrike Ehlert, and Dirk H. Hellhammer, "Effects of Suckling on Hypothalamic-Pituitary-Adrenal Axis Responses to Psychosocial Stress in Postpartum Lactating Women," *Journal of Clinical Endocrinology & Metabolism*, October 1, 2001, vol. 86, no. 10, pp. 4798–4804.

Religious faith and suicide risk examined: Ryan E. Lawrence, Maria A. Oquendo, and Barbara Stanley, "Religion and Suicide Risk: a systematic review," *Archives of Suicide Research*, vol. 20, no. 1, 2016, pp. 1–21. The researchers believe that religious faith does not reduce suicidal thoughts but that it reduces suicide attempts.

Seasonality of suicide risk: Marc Shapiro, "Suicide Rates Spike in Spring, Not Winter," May 8, 2019, *Johns Hopkins Medicine*, May–June 2019 issue, https://www.hopkinsmedicine.org/news/articles/suicide-rates-spike-in-spring-not-winter. See also Sally C. Curtin, Holly Hedegaard, and Farida B. Ahmad, "Provisional Numbers and Rates of Suicide by Month and Demographic Characteristics: United States, 2020," National Vital Statistics, November 3, 2021, Centers for Disease Control.

Relationship between religiosity and suicide: Ning Hsieh, "A Global Perspective on Religious Participation and Suicide," *Journal of Health and Social Behavior*, June 22, 2017, vol. 58, no. 3, https://doi.org/10.1177/0022146517715896.

Baleen whales less likely to engage in mass strandings: A. S. Aniceto, L. Tassara, A. Rikardsen, and P. Blévin, "Mass strandings of seven toothed and baleen whale species in Northern Norway in March 2020 call for further investigation," *Polar Biology*, May 6, 2021, vol. 44, pp. 1457–1461.

Canine suicide: Stanley Coren, "Do Dogs Commit Suicide?" *Psychology Today*, August 23, 2010.

Male and female dolphin testosterone levels: Vicky Kirby, Joan Sawyer-Steffan, and W. G. Gilmartin, "Hormones and Sexual Activity in the Atlantic Bottlenosed Dolphin (*Tursiops truncatus*)," IAAAM Archive, Naval Oceans Systems Center, https://www.vin.com/apputil/content/defaultadv1.aspx?id=3863396&pid=11257&. See also Kathryn K. Sherman, Nicole E. Beaulieu-McCoy, Elyse L. Wurster, Randall S. Wells, Cynthia R. Smith, Aaron A. Barleycorn, Jason B. Allen, and Nicholas M. Kellar, "Serum correlation, demographic differentiation, and seasonality of blubber testosterone in common bottlenose dolphins, *Tursiops truncatus*, in Sarasota Bay, FL," *Scientific Reports*, April 26, 2021, vol. 11. no. 1, https://pubmed.ncbi.nlm.nih.gov/33903714/.

Theory proposed for collective dolphin self-destruction: Braitman, *Animal Madness*, pp. 173–174.

Relationship between eating disorders and obsessive-compulsive disorder: Katherine A. Phillips and Walter H. Kaye, "Relationship of Body Dysmorphic Disorder and Eating Disorders to Obsessive-Compulsive Disorders," in *Obsessive-Compulsive Spectrum Disorders: Refining the Research Agenda for DSM-V*, edited by Eric Hollander, Joseph Zohar, Paul J. Sirovatka, and Darrel A. Regier, American Psychiatric Association, Arlington, Virginia, 2011. See also Howard Steiger, "Eating disorders and the serotonin connection: state, trait and developmental effects," *Journal of Psychiatry & Neuroscience*, January 2004, vol. 29, no. 1, pp. 20–29.

CHAPTER 14: "WHY PRIMATOLOGISTS MONKEY AROUND"

Similarities and differences in facial expression of humans and apes: Annukka K. Lindell, "Continuities in emotion lateralization in human and non-human primates," *Frontiers in Human Neuroscience*, August 7, 2013, https://doi.org/10.3389/fnhum.2013.00464. See also William D. Hopkins, "Comparative and Familial Analysis of Handedness in Great Apes," *Psychological Bulletin Journal*, July 2006, vol. 132, no. 4, pp. 538–559. See also Lisa A. Parr, "Understanding the Expression and Classification of Chimpanzee Facial Expressions," in *The Mind of the Chimpanzee: Ecological and Experimental Perspectives*, edited by Elizabeth V. Lonsdorf, Stephen R. Ross, and Tetsuro Matsuzawa, University of Chicago Press, Chicago, 2010, pp. 50–59. See also Ethan G. Harrod, Christopher L. Coe, and Paula M. Niedenthal, "Social Structure Predicts Eye Contact Tolerance in Nonhuman Primates: Evidence from a Crowd-Sourcing Approach," *Nature: Scientific Reports*, April 4, 2020, vol. 10, no. 6971. See also Anne M. Burrows, Bridget M. Waller, Lisa A. Parr, and Christopher J. Bonar, "Muscles of facial expression in the chimpanzee (*Pan troglodytes*): descriptive, comparative and phylogenetic contexts," *Journal of Anatomy*, February 2006, vol. 2, no. 208, pp. 153–167.

Noted animal trainer's view of chimps and gorillas: Karen Pryor, *Reaching the Animal Mind: Clicker Training and What It Teaches Us About All Animals*, Scribner, New York, 2009, pp. 84 and 132.

CHAPTER 15: "MAIN LESSONS OF EVOLUTION—INCLUDING HOW AND WHY WE HAVE MISUNDERSTOOD THE CAUSES OF WAR AND PEACE"

Change in public attitudes toward Napoleon from the beginning of the 100 Days Campaign to Waterloo: Philip Mansel, *Paris Between Empires: Monarch and Revolution, 1814–1852*, St. Martin's Press, New York, 2014, pp. 66–75.

Richard Dawkins's claims about human altruism: Richard Dawkins, *The Selfish Gene*, Oxford University Press, Oxford, UK, 1976, p. 176.

Richard Dawkins's claim that religious parents should have their children taken away from them: Jenn Selby, "Richard Dawkins: 'Children need to be protected' from religious parents," *The Independent*, February 26, 2015.

ACKNOWLEDGMENTS

I OWE THANKS TO MANY people for this book. But there is one who stands far above all the others in respect of my debt of gratitude: my wife, Christina Freyss. Christina read over the manuscript and gave me thoughtful advice about it. She also examined all of its pictures and graphs. She is a national award-winning graphic designer, and she reformatted the images, redesigned the charts, and even designed the cover.

I am also grateful to two people at the Emory National Primate Research Center: its director, R. Paul Johnson, and the director of the Laboratory for Darwinian Neuroscience, James Rilling. I'm grateful for their time and assistance. Please permit me to add this point about the Primate Research Center: While this book is critical of some of its prior claims about the degree of kinship between humans and apes, I hope that readers understand the enormous importance of the work that it does on brain research and on primate immune response. That latter work is carried on as well at other national primate research labs, and this may prove to be vitally important in a future pandemic. Legislators responsible for determining the amount of its funding should bear this in mind.

I am grateful as well to David DesRosiers for giving me permission to use a portion of an article I wrote that originally appeared on the website Planned Manhood. Needless to say, too, I'm grateful to my publisher, David Bernstein, my Post Hill Press managing editor, Aleigha Koss, and to the company's copy editor Donna DuVall. David had a number of tremendously helpful suggestions, Aleigha skillfully kept everything moving along, and Donna did an extraordinary job of copyediting a manuscript with innumerable knotty points regarding grammar and spelling, along with hundreds of source notes. My thanks, as well, to John Craig, who read a very early draft of the book. I'm also indebted to my mother and my now deceased stepmother and father, each of whom informed me in different ways about neuroscience and neuropsychology.

There are many others who have contributed to this book in ways large and small, and I hope they do not feel slighted that I have not mentioned them by name.

ABOUT THE AUTHOR

By Christina Freyss

JONATHAN LEAF IS A PLAY-WRIGHT and novelist. He has written for the *Daily Beast*, *Commentary*, *National Review*, *The New Criterion*, *Modern Age*, *Tablet*, the *New York Post*, *New York Daily News*, *Spectator* (USA), *Law & Liberty*, *City Journal*, *The Weekly Standard*, and many other publications.

In 2018, *The Wall Street Journal* called his play *Pushkin* a "triumph," naming it one of the year's four best. *Kirkus Reviews* has called his novel *City of Angles* "literary entertainment at its best." Now Leaf turns his attention to the subject that his parents devoted their lives to and with which he has had a lifelong fascination: what neuroscience is revealing about human nature.

INDEX